HOW TO GROW
A HUMAN

HOW TO GROW A HUMAN

ADVENTURES IN HOW WE ARE MADE AND WHO WE ARE

PHILIP BALL

THE UNIVERSITY OF CHICAGO PRESS

The University of Chicago Press, Chicago 60637
© 2019 by Philip Ball
All rights reserved. No part of this book may be used or reproduced in any
manner whatsoever without written permission, except in the case of brief
quotations in critical articles and reviews. For more information, contact
the University of Chicago Press, 1427 E. 60th St., Chicago, IL 60637.
Published 2019
Printed in the United States of America

28 27 26 25 24 23 22 21 20 19   1 2 3 4 5

ISBN-13: 978-0-226-65480-5 (cloth)
ISBN-13: 978-0-226-67617-3 (e-book)
DOI: https://doi.org/10.7208/chicago/9780226676173.001.0001

First published in Great Britain by William Collins, an imprint of
HarperCollins Publishers, 2019.

LIBRARY OF CONGRESS CATALOGING-IN-PUBLICATION DATA
Names: Ball, Philip, 1962– author.
Title: How to grow a human : adventures in how we are made and
who we are / Philip Ball.
Description: Chicago : The University of Chicago Press, 2019.
| Includes bibliographical references and index.
Identifiers: LCCN 2019008607 | ISBN 9780226654805 (cloth : alk. paper) |
ISBN 9780226676173 (e-book)
Subjects: LCSH: Tissue engineering—Popular works. | Tissue culture—
Popular works. | Organ culture—Popular works. | Cell transformation—
Popular works. | Cytology—Popular works. | Developmental biology—Popu-
lar works. | Bioengineering—Popular works.
Classification: LCC R857.T55 B35 2019 | DDC 612/.028—dc23
LC record available at https://lccn.loc.gov/2019008607

♾ This paper meets the requirements of ANSI/NISO Z39.48-1992 (Permanence
of Paper).

CONTENTS

PROLOGUE

This book arose from an astonishing experiment in which I was invited to take part, and which involved the transformation of a piece of me.

In the course of writing it, as I spoke to scientists about cell biology and fertility, embryology and medicine, philosophy and ethics, it dawned on me just how much science, in these fields especially, is driven by stories.

I don't mean "human stories" – those tales about people that are said to be so central to keeping general audiences engaged. I mean that our perception of what science *means* is shaped by narratives about it. There are narratives that society imposes on new discoveries and advances, which are often cut from older cloth. In biology – particularly developmental, cell and reproductive biology – these stories tend to come from myth, science fiction and fantasy, and they are often alarming: they might, for example, draw on *Frankenstein, Brave New World, The Island of Doctor Moreau.* But there are also narratives that scientists themselves create and recycle. It happens more in the biological sciences than in the others because biology is inherently a science of becoming, where history matters and where we seem compelled to speak about goals and purposes: what organisms, cells and genes "want", what evolution "seeks to do".

A significant aspect of this book's aim is to expose and explore these stories. They are by no means a bad thing. On the contrary, they are essential and often illuminating, for they fit with the way we humans make sense of our world, with our instinct to look for and find causation, reasons behind things. Yet there is a danger to them too, which is that we might begin to mistake them for descriptions of how things are.

One of the classic examples of a framing narrative in biology is Richard Dawkins's concept of the selfish gene. Dawkins has had to defend this idea against accusations that it attributes a kind of agency and intention to genes, for it is of course just a metaphor. The problem is that, when a metaphor gets this popular, it starts to be understood (and sometimes presented) as a plain account of "what things are like". When Dawkins speaks of us – individual humans – as the "survival machines" for genes, he is not defining what a human is but is explaining what we are required to be within the narrative of the selfish gene. The narrative serves to convey a particular aspect of how genes function in evolutionary terms. If you don't like this story, or don't find it helpful, you are not obliged to accept it; there is nothing "real" about it. Dawkins has said as much: he admits in *The Selfish Gene* that it should strictly be called *The Slightly Selfish Big Bit of Chromosome and the Even More Selfish Little Bit of Chromosome*. (He was right to suspect the book would not have been so successful if he'd called it that.) He even admitted much later that the book could equally have been titled *The Cooperative Gene*. That would have been a different narrative, of course, and it would have served to illustrate another facet of how genes work.

As Dawkins implicitly acknowledges here, biology is too complex and complicated to be reduced to a single story. That's precisely why we *need* stories to tell about it: they give us something to cling to, some way of finding a path through the thicket. There is almost always more than one of them.

It's not just, though, that we must remember this is what we are doing. We must also keep in mind that stories are not neutral vehicles for understanding. When we frame some medical advance within the narrative of *Brave New World*, we are not simply saying, "hey, doesn't this sound a bit like the people-growing hatcheries in Aldous Huxley's book!", but also " . . . and we should be suspicious, even frightened, of where it might lead." Precisely the same applies to *The Selfish Gene*. The story's subtext here is that Darwinian evolution is ruthless – that it makes for a dog-eat-dog world, a battle for survival from the bottom up. Dawkins explains that this does not imply that humans are themselves bound to be selfish; indeed, he shows how altruism can arise from "selfish genes". But the implication remains that there is to nature a redness in tooth and claw, and indeed Dawkins advises that for this very reason we should strive to supersede the default and be kind to one another. The point is, though, that one can talk about Darwinian evolution and genes without invoking "selfishness" at all, and the story then has a different complexion in which all manner of behaviours appear not as emergent and perhaps counter-intuitive consequences of a selfish genetic strategy but simply as aspects of the complexity of biology: cooperation as well as the most beastly predation strategies, war and peace, love and cruelty. Each of those words is equally freighted with narratives that biology itself doesn't impose.

This is why I will be constantly alert to the narrative, and will ask, "why this story, and not some other?" Whether we are talking about cancer or immunity, cell signalling or tissue engineering, the science becomes packaged right from the outset inside a story, and this means that we impute agency, make choices about what to include and what not to, and suggest certain goals and not others. Even scientists speaking to other scientists need to use metaphors and narratives somewhere along the road, to give the mind purchase on concepts that are otherwise too slippery and too complex to comprehend. The only danger in all this, as in stories about "selfish genes", is if we tell ourselves that all we are doing is relating objective truths.

I'd suggest that you might want to be alert to the narratives that I shall deploy too – for I am no more immune than anyone else to the tendency, the need, to tell a story, and the habit of using framing devices unconsciously. Challenge me on it. I promise to try not to mind.

It's for this reason too that I think it is always important to know something about the historical context in which a scientific idea has arisen. We will see that, for example, cell theory was originally deemed to have a political dimension, and tissue culture was driven by social agendas. Some scientists might say, "oh, but that was then, and we have shed that baggage now and are dealing just with plain facts." But I suspect that few scientists working on fertility and infertility would say this, and certainly they should not. They know very well that whatever they discover will be refracted through a complex social legacy of attitudes to baby-making, sex and gender. Geneticists labour under the shadow of the field's eugenic past, and this goes beyond the stark fact that such work once led to enforced sterilization of the "unfit" in many countries and was embraced by the Nazis. Unease and dispute remain today over the implications of genetics for incendiary issues like race, class, intelligence and disability. What this means is that culture, past and present, may shape the scientific questions we ask, the models we develop, and the stories we tell.

I know from experience that there is a kind of reader who says,

"I don't care about that, just give me the science!" If you are that kind of reader, I'd humbly reply: I cannot give you "just the science", because it already comes with a story attached. In this field – which I find breathtaking, perplexing, occasionally disturbing – there is never "just the science".

When we ask, "how to grow a human", we cannot possibly be asking "just" a question about science. That's what makes the question so interesting.

INTRODUCTION

MY BRAIN IN A DISH

In the summer of 2017, a small piece of my arm was removed and turned into a rudimentary miniature brain. This book is my attempt to make sense of that strange experience.

On a hot day that July, I lay on a bed in the Institute of Neurology at University College London while neuroscientist Ross Paterson gouged a little chunk from my shoulder with a kind of miniature surgical apple-corer. A dab of local anaesthetic made it painless; to my great relief, there wasn't much blood.

Bathed in a nutrient solution in a test-tube, that piece of my flesh was the seed for what, eight months later, would resemble a tiny brain.

It was my own "mini-brain", a blob of neurons about the size of a lentil. They wired themselves into a dense network and could signal to one another in the way neurons do. I'm not going to say it was thinking; probably these signals were not much more than random sparks, incoherent noise, signifying nothing. But no one really knows how to think about what goes on inside a mini-brain, any more than they know what transpires in the formative brain of a fetus when it is of comparable pea-size.

This process of culturing new tissues from a piece of arm is not the way to create a human. But it could one day become a basis for doing that.

It's not obvious why anyone, here and now, would think this a good idea. But the point is not that one day humans will, like the citizens of Aldous Huxley's *Brave New World*, be cultivated from blobs of cells in the vats of some dystopian people-factory. The point is that such a vision is no longer obviously impossible. That alone should give us cause to revise our ideas about what we think we are. Having a piece of you grown into a mini-brain in an incubator five miles across town brings home to you, rather viscerally, why the need for that revision is upon us now.

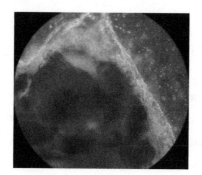

My mini-brain under the microscope.

∗ ∗ ∗

Let me explain.

My mini-brain was cultivated, nurtured and guided by neuroscientists Selina Wray and Christopher Lovejoy at UCL. That's their job. They want to understand how brains develop, and in particular why some gene mutations send that process awry and trigger the onset of neurodegenerative diseases such as Alzheimer's. These conditions, which many of us will confront – close to a million people in the UK are currently thought to be living with some form of dementia – are partly a consequence of ageing, but they can also have genetic roots. Some gene mutations confer particular susceptibility to dementias, and there are inheritable early-onset dementias that can affect people even in their thirties. My "brain in a dish" was grown as part of a large and ambitious project called Created Out of Mind, funded in 2016–18 by the Wellcome Trust to alter public perceptions of dementia and develop new tools for assessing the value of arts-based interventions for people living with these conditions.

Selina and Chris hope that, by studying the activity of genes in mini-brains cultured from the tissues of people with those genetic mutations, they might come to understand more about the causes, and ultimately find clues that could lead to possible cures. Scientists studying the genetic factors behind the neurodegenerative Huntington's disease have already found that the baleful effect of one particular gene implicated in the condition can be tempered with drugs that intercept the conversion of the gene to a protein that is prone to "misfolding". It's this misfolded form that produces scarring and destruction of brain tissue. Others are working on vaccines that might prevent or remove the clumps of misfolded protein in the brain that seem to trigger Alzheimer's itself.

To my knowledge, I do not have genes that make me susceptible to early-onset Alzheimer's. But the aim of the Brains In A Dish project, coordinated by artist Charlie Murphy, was to explore and explain research like this through the response of its participants. Well, this book is mine.

I should be very clear what this term "mini-brain" implies. Some researchers reject it, and I see their point. Human neurons grown in a cell culture in this way can't make a brain, not even in its early fetal form. But these nerve cells do start to create, under the direction of their own genetic programme, some of the features that a real developing brain exhibits. They become specialized into some of the many different cell types – not just neurons – that are found in our mature brains. And they acquire some of the anatomical structure of brains: the well-defined layers of neurons seen in the cortex, the folds and convolutions of the tissues. It's rather like a very young child's drawing of a person: not much of a resemblance really, but you can see what they're getting at. You can see the potential to do a better job. A more neutral term for these lab-grown cell structures is "organoid": the cells construct something that looks like a crude representation of an organ of the body, reduced in scale. It is possible to grow organoids resembling livers, kidneys, retinas, gut, as well as brains – all in a dish, outside the body. I want to ask what this means, for medicine, fundamental biology, philosophy and our sense of identity.

There was no rulebook to tell me how I should feel about my mini-brain. Certainly, I didn't lie awake at night fretting over its welfare; this mass of tissue made from my skin didn't take on the status of an individual. But I felt oddly fond of those cells, doing their best to fulfil a role in the absence of the guiding influence of their somatic source.* There was a curious intimacy involved, a sense of potential that wasn't present initially in the tiny chunk of arm-flesh excised and placed in a test-tube. This was more than a matter of cells subsisting; this was life in all its teeming, multiplying glory, spilling out from a paring of me.

------

* "Somatic" is a word we will encounter a lot. It simply means "bodily" or "of the body".

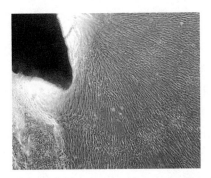

My skin cells (fibroblasts) growing in a petri dish (in vitro)
from a piece of skin tissue taken from my arm.

It's hard not to invest the whole of biology with intentions, purposes, wants and needs, even though cells and simple organisms resemble automata responding without volition to the signals from their environment. (Some would say this applies to humans too.) That's just how nature works.

Yet seeing these cells do their business in a petri dish is to recognize that life exists in doing. It is a process in which change is the only constant: change imbued with direction, more or less constrained onto a trajectory that evolution has guided and given what looks almost indistinguishable from a purpose, until death makes that change an inexorable slide into decay and entropic dissolution. There is no agreed scientific definition of "life", but such a thing (if it is possible at all) would mean little if it fails to acknowledge this dynamic aspect, this interplay of predetermined pattern and historical contingency. It feels banal to say that my excised cells took on a life of their own, but what is so new and so remarkable about the science behind organoids is that we have the knowledge and power to influence the direction that life takes.

* * *

What *really* needs explaining is that my mini-brain was grown from a piece of my *arm*: in essence, from skin cells. That doesn't sound like something that should be possible. Until barely more than a decade ago, most biologists thought so too. The discoveries that changed this view have transformed cell biology, raising all manner of medical possibilities for regenerating organs and tissues as well as opening up new avenues for basic research into embryology, development and conception. These discoveries and their applications are at the heart of this book.

But although these cell-transforming technologies have been celebrated, sometimes in breathless and hyperbolic terms, in the popular press, I'm not at all sure that their wider philosophical, one might even say psychic, implications have been acknowledged. Here is one of the profound things they say to us:

Every part of ourselves can potentially be turned into any other part of ourselves – including a complete self.

This, let me add, is more than science has yet proved. There is some small print attached, and further ingredients might be needed to fulfil the last part of the bargain. All the same, we are more plastic than we ever guessed. And that realization is in turn a culmination of medical advances and discoveries that have taken place over the course of the past century, which too we have processed only in the manner that we always process discoveries we don't know how to think about. That's to say, we have framed it around fears, fantasies and fiction.

For example: perhaps my "brain in a dish" immediately invokes visions of *Frankenstein*. And I pointed you also towards *Brave New World*, that re-envisioning of Mary Shelley's tale for the age of industrialization and mass culture. These two novels remain today the favourite, off-the-shelf points of cultural reference for biomedical advances that unsettle and boggle the mind. But we are sent back to other speculative fictions by some of the possibilities that I have seen seriously and soberly discussed in the context of cell transformation and organoid growth. For example:

- It may well be possible to grow a human brain in the body of a pig. One reason I find this so disturbing is that I still vividly remember first seeing the scene in Lindsay Anderson's 1973 film, *O Lucky Man!*, where . . . well, if you don't know it already, don't let me spoil it for you.
- It might be feasible to grow each organ of the human body separately outside the body itself in some sort of vessel (*in vitro*), and then surgically assemble them into a person – or enough of a person to be, let's say, a personoid. And this is precisely how the first robots were made, in Karel Čapek's 1921 play, *R.U.R.*
- Philosophers, ethicists and neuroscientists are now compelled to debate what it could mean to create a full-grown human brain in a vat. Could it be conscious? Would it experience a self-contained interior "reality"? The pop-culture reference point here, which philosophers have embraced with nerdy delight, is of course the Matrix movies of the Wachowskis.

Let me make it clear that no researcher sees any prospect of these things happening in the near future, nor any good reason to try to make them happen. I will look at them more closely later, but my point here is not to tell rather shocking and thrillingly grotesque scare stories in a bout of bait-and-switch. What matters is that, confronted with so disorienting and disturbing a set of *imaginable* possibilities, we seem to *need* such stories in order to frame our thoughts. That in itself is worth considering. What motivates and shapes these narratives?

Underpinning them, I believe, is a consideration that might at first strike you as odd: *We are not at ease with our own flesh.*

But surely, you might say, we *inhabit* our own flesh? I'm putting those words into your mouth because they have a familiar ring to them – it doesn't seem a strange thing to suggest. Yet such a phrase serves more to disassociate than to unite us with our flesh. We inhabit it? Like a person who inhabits a house? So what, then, are "we"? It's the old Cartesian dualism: the separation of mind and body, or as some might have once said, of body and soul.

Yet of course we are not at ease in our own flesh! How we recoil from its routine functions, its excrescences and smells and fluids. How hard we try to remodel it, with what horror we watch its decrepitude. How we flock to watch movies like *Hellraiser* and *Saw*, or, at the more sophisticated end of the body-horror market, pretty much any of David Cronenberg's early oeuvre. Gunther von Hagens, with his plasticized corpses, has made a career from an artful exploration of our horrified fascination; artists from Mark Quinn (who has sculpted with his own blood and the placenta of his baby son) to Marina Abramović have made what are often brave, painful and stomach-churning attempts to help us engage with our raw material.

There are many reasons for this ambivalence towards the somatic aspects of human existence, expressed repeatedly in all cultures in all ages: with piercings and tattoos, rituals of embalming and burial, elaborate taboos and the normative strictures of surgery. But the new sciences of culturing and transforming our cellular fabric confront us with perhaps the most fundamental challenge for our relationship with our flesh. They show us the ultimate in dismemberment: a reduction of the person to the cell.

Time was when we could dismiss that. Sure, cells are our building blocks, but no more or less so than proteins, atoms or quarks. If a chunk of our cellular material were removed, well, so what? It was no longer "us", but a piece of waste, separate and dead, soon to be putrescent and doomed by microbes and entropy.

Having your skin grown into neurons that assemble themselves into a brain organoid is a pretty convincing way to find out how obsolete that notion is. All the more so when you see it through the microscope and realize that this is not merely some trick of preservation. Life on that scale is multitudinous, and thriving, and it has a plan of sorts.

Life? Whose life?

Not mine, exactly – and yet who else's can it be? Those cells are autonomous – but why any more so than the cells still in my arm, in my real brain (the need for that specifying adjective still makes me blink), my surging blood and beating heart? And so I come by degrees to accept the inevitable truth: I am a colony of cells, whose cooperation lets me draw breath, whose communication produces my sense of identity and uniqueness.

That is what is fundamentally disconcerting about our own flesh. It grew as a colony from a single cell, and we're not quite sure where (or when) in this teeming morass to pin the label "me".

The new cell technologies are making it impossible for us any longer to ignore this fact. I won't pretend that I know how to normalize it, but I think there is a strange kind of liberation that can come from letting ourselves be unsettled by it.

CHAPTER 1

PIECES OF LIFE

CELLS PAST AND PRESENT

"*Ex ovo omnia.*"

So declared the frontispiece of *Exercitationes de generatione animalium* (1651) by the seventeenth-century English physician William Harvey, sometime physician to James I. It expressed a conviction (and no more) that all things living come from an egg.

William Harvey's motto "All things from an egg" in the frontispiece of his 1651 treatise on the "generation of animals".

It's not really true: plenty of living organisms, such as bacteria and fungi, do not begin this way. But we do. (At least, we have done so far. I no longer take it for granted that this will always be the case.)

"Egg" is an odd term for our generative particle, and indeed Harvey was a little vague about what he meant by it. Strictly speaking, an egg is just the vessel that contains the fertilized cell, the *zygote* in which the male genes from sperm are combined with the female genes from the "egg cell" or ovum. Yet it's easy to overlook how bold a proposal Harvey was making, in a time when no one (himself included) had ever seen a human ovum and the notion that people might begin in a process akin to that of birds and amphibians could have sounded bizarre.

The truth of Harvey's insight could only be discerned once biology acquired the idea of the cell, the fundamental "atom of biology". That insight is often attributed to Harvey's compatriot and near-contemporary Robert Hooke, who made the most productive use of the newly invented microscope in the 1660s and '70s. Hooke discerned that a thin slice of cork was composed of tiny compartments that he called "cells". This is often said to be an allusion to the cloistered chambers (Latin *cella*: small room) of monks, but Hooke drew a parallel instead with the chambers of the bee's honeycomb, which in turn probably derive from the monastic analogy.

Robert Hooke's sketch of cells in cork, as seen in the microscope.

The popular notion that Hooke established the cellular basis of all living things is wrong, however. Hooke saw cells, for sure – but he had no reason to suppose that the fabric of cork had anything in common with human flesh. In fact, Hooke imagined that cork cells are mere passages for transporting fluids around the cork tree. The cell here is just a passive void in the plant's fabric, a very different notion from the modern concept of an entity filled with the molecular machinery of life.

More provocative were the observations of the Dutch cloth merchant Antonie van Leeuwenhoek in 1673 that living organisms can be of microscopic size. Leeuwenhoek saw such "animalcules" teeming in rainwater – mostly single-celled organisms now called protists, which are bigger than most bacteria. You can imagine how unsettling it was to realize that the water we drink is full of these beings – although not nearly so much so as the dawning realization that bacteria and other invisibly small organisms are everywhere: on our food, in the air, on our skin, in our guts.

Leeuwenhoek contributed to that latter perception when he discovered animalcules in sperm. That was one of the substances that the secretary of London's Royal Society, Henry Oldenburg, suggested the Dutchman study after receiving his communication on rainwater. Leeuwenhoek looked at the semen of dogs, rabbits and men – including his own – and observed tadpole-like entities that "moved forward with a snake like motion of the tail, as eels do when swimming in water." Were these parasitic worms? Or might they be the very generative seeds themselves? They were, after all, absent in the sperm of males lacking the ability to procreate: young boys and very old men.

Here we can see the recurrent tendency to impose familiar human characteristics on what is evidently "of us" but not "like us". The French physician Nicolas Andry de Boisregard, an expert on tiny parasites and a microscope enthusiast, claimed in 1701 that "spermatic worms" could be considered to possess the formative shape of the fetus: a head with a tail. In 1694, the Dutch microscopist Nicolaas Hartsoeker drew a now iconic image of a tiny fetal humanoid with a huge head tucked up inside the head of a spermatozoa: not, as sometimes claimed, a reproduction of what he thought he could see, but a figurative representation of what he imagined to be there.

The "homunculus" in sperm, as imagined by Nicolaas Hartsoeker.

This was one of the most explicit expressions of the so-called preformationist theory of human development, according to which the human body was fully formed from the very beginning of conception and merely expands in size: an extrapolation down to the smallest scale of the infant's growth to adulthood. According to this picture, the female egg postulated by Harvey continued to be regarded in the prejudiced way in which Aristotle had perceived the woman's essentially passive role in procreation: it was just a receptacle for the homunculus supplied by the man.

That was, though, a different view to the one Harvey envisaged, in which the body developed from the initially unstructured egg. Harvey, like Aristotle, thought that semen triggered this process of emergence, which Aristotle had imagined as a kind of curdling of a fluid within the female. The idea that the embryo unfolds in this way, rather than simply expands from a preformed homunculus, was known as *epigenesis*. These rival views of embryo formation contended until studies with the microscope, particularly investigations of the relatively accessible development of chicks inside the egg, during the eighteenth and early nineteenth century put paid to the preformation hypothesis. Embryos gradually develop their features, and the question for embryologists was (and still is) how and why this structuring of the tissues comes about.

* * *

The observations of early microscopists did not, then, engender a belief that life is fundamentally cellular. That cells are a general component of living matter was not proposed until the early nineteenth century, when the German zoologist Theodor Schwann put the idea forward. "There is one universal principle of development for the elementary parts of organisms," he wrote in 1839, "and this principle is in the formation of cells."

Schwann developed these ideas while working in Berlin under the guidance of physiologist Johannes Müller. One of Schwann's colleagues in Müller's laboratory was Matthias Jakob Schleiden, with whom he collaborated on the development of the cell theory. Schleiden's prime interest was plants, which were more easily seen under the microscope to have tissues made from a patchwork of cells, their walls constituting emphatic physical boundaries. This structure wasn't always evident for animal tissues (especially hair and teeth), but Schwann and Schleiden were convinced that cell theory could offer a unified view of all living things.

Schleiden believed that cells were generated spontaneously within the organisms – an echo of the notion of "spontaneous generation" of living matter that many scientists still accepted in the early nineteenth century. But he was shown to be wrong by another of

Müller's students, Robert Remak, who showed that cells proliferate by dividing. Remak's discovery was popularized – without attribution – by yet another Müller protégé, Rudolf Virchow, who tends now to be given the credit for it. All cells, Virchow concluded, arise from other cells: as he put it in Harveian manner, *omnis cellula e cellula*. New cells are created from the division of existing ones, and they grow between successive divisions so that this series of splittings doesn't result in ever tinier compartments. Virchow proposed that all disease is manifested as an alteration of cells themselves.

Virchow was the kind of person only the nineteenth century could have produced – and perhaps indeed only then in Germany, with its notion of *Bildung*, a cultivation of the intellect that encouraged the emergence of polymaths like Goethe and Alexander von Humboldt. Virchow studied theology before taking up medicine in Berlin. While establishing himself as a leading pathologist and physician, he also became a political activist and writer and was involved in the uprisings of 1848. As if to demonstrate that nothing was simple in those times, this eminent biologist and religious agnostic was also profoundly opposed to Charles Darwin's theory of evolution, of which his student Ernst Haeckel was Germany's foremost champion.

Virchow thus had fingers in many pies; but in his view they were different slices of the same pie. The influence of politics, ideology and philosophy on science is always clearer in retrospect, and that's nowhere more apparent than in the physiology of the nineteenth century. For Schwann, "each cell is, within certain limits, an Individual, an independent Whole": an idea indebted to the Enlightenment celebration of the individual. The cell was a living thing – as the physiologist Ernst von Brücke put it in 1861, an "elementary organism" – which meant that higher organisms were a kind of community, a collaboration of so many autonomous, microscopic lives, in a manner that paralleled the popular notion of the nation state as the collective action of its citizens. Meanwhile, Schwann's conviction of the cellular nature of all life, implying a shared structural basis for both plants and animals, was motivated by his sympathy for the German Romantic philosophical tradition that sought for universal explanations.

For Virchow, this belief in tissues and organisms as cellular collectives was more than metaphor. It was the expression writ small of a principle that applied to politics and society. He was convinced that a healthy society was one in which each individual life depended on the others and which had no need of centralized control. "A cell . . . yes, that is really a person, and in truth a busy, an active person," he wrote in 1885. "What the individual is on a grand scale, the cell is that and perhaps even more on a small one."

Life itself, then, showed for Virchow how otiose and mistaken was the centrist doctrine of the Prussian statesman Otto von Bismarck, who at that time was working towards the unification of the German states. Virchow attacked Bismarck's policies at every opportunity and denounced his militaristic tendencies, so enraging the German nobleman that Bismarck challenged Virchow to a duel. The physiologist shrugged it off, making Bismarck's belligerent bluster seem like the aristocratic posturing of a bygone era.

The idea of the "germ" as a rogue microbial invader was the counterpart to the idea of a body as a community of collaborating cells. It was entirely in keeping with the political connotations of cell theory that germ theory blossomed in parallel. Once Louis Pasteur and Robert Koch showed in the nineteenth century that micro-organisms like bacteria can be agents of disease ("germs"), generations of children were taught to fear them. Germs are everywhere, our implacable enemies: it is "man against germs", as the title of a 1959 popular book on microbiology put it. After all, hadn't a bacillus been identified as the cause of cholera in

1854? Hadn't Pasteur and Koch established bacteria as the culprits behind anthrax, tuberculosis, typhoid, rabies? These germs were nasty agents of death, to be eradicated with a thorough scrub of carbolic. And to be sure, the antiseptic routines introduced by the unjustly ridiculed Hungarian-German physician Ignaz Semmelweis in the 1840s (as if washing your hands before surgery could make any difference!), and later by Joseph Lister in England, saved countless lives.

This new view of disease had profound sociopolitical implications. The old notion of a disease-generating "miasma" – a kind of cloud of bad air – situated illness in a particular locality. But once disease became linked to contagion passed on between *people*, a different concept of responsibility and blame was established. The politicized and racialized moral framework for germ theory is very clear in the description of a French writer from 1885 who talked of disease as "coming from outside, penetrating the organism like a horde of Sudanese, ravaging it for the right of invasion and conquest." This was the language of imperialism and colonialism, and disease was often portrayed as something dangerously exotic, coming from beyond the borders to threaten civilization. Those who supported contagion theory tended to be politically conservative; liberals were more sceptical.

From the outset, then, our cellular nature was perceived to entail a particular moral, political and philosophical view of the world and of our place within it.

* * *

We come from a union of special cells: the so-called gametes of our biological parents, a sperm and an egg cell. Perhaps because this fact is taught in primary schools – a reassuringly abstract vision of human procreation, typically presented with no hint of the confusing richness and peculiarity of stratagems and diversions it engenders – we forget to be astonished by it. That we begin as a microscopic next-to-nothing, a pinprick somehow programmed with potential, *is* remarkable and counter-intuitive. It's slightly absurd to image that infants will casually accommodate this "fact of life", which seems on reflection barely conceivable. Only their endless capacity to accept the magical with equanimity lets us get away with it.

And so it continues through the educational grades: cells exist, and that is all we need to know. School children are taught to label their parts: the oddly named bits like mitochondria, vacuoles, endoplasmic reticulum, the Golgi apparatus. What has all this to do with arms and legs, hearts, brains? One thing is sure: the cell is anything but a homunculus. In which case, where does the body come from?

What the cell and the body do have in common is *organization*. They aren't random structures; there's a plan.

But that word, so often bandied about in biology, is dangerous. It doesn't seem likely that we can ever leach from the notion of a plan its connotations of foresight and purpose. To speak, as biologists do, of an organism's "body plan" is to lay the ground for the idea that there exists somewhere – in the cell, for where else could it be? – a blueprint for it. But it is not merely by analogy, but rather as a loose parallel, that I would invite you then to wonder where exists the plan, the blueprint, for a snowflake. There's a crucial distinction between living organisms and snowflakes, and it is this distinction that explains why snowflakes do not evolve one from the other via inheritance but are created *de novo* from the cold and humid winter sky.* But there are good motivations for the

---

* Some crystals, and parts of crystals, do "inherit" structure from others, thanks to their ability to act as a kind of template for further growth. It is for this reason that one speculative theory for the origin of life has proposed that clay minerals were able to act as a kind of prebiotic, inorganic carrier of genetic information,

comparison, for the body's organization, like the snowflake's, is an instantiation of a particular set of rules that govern growth.

Virchow and his contemporaries were already coming to appreciate that cells are more than fluid-filled sacs. In 1831, the Scottish botanist Robert Brown reported that plant cells have a dense internal compartment that he called a nucleus.* By Virchow's time, cells were considered to have at least three components: an enclosing membrane, a nucleus, and a viscous internal fluid that the Swiss physiologist Albert von Kölliker named the cytoplasm. (*Cyto-* or *cyte* means cell, and we'll see this prefix or suffix a lot.)

Kölliker was one of the first to study cells microscopically using the technique of staining: treating them with dyes that are absorbed by cell components to make their fine structure more visible. He was a pioneer in the field of *histology*, the study of the anatomy of tissues and their cells. Kölliker was particularly interested in muscle cells, and he showed that these are of more than one type. One variety has a stripy ("striated") appearance when stained, and Kölliker noticed that striated muscle cells also contain many tiny granules, later identified as another component of animal cells that were named mitochondria in 1898. Around the same time, other substructures of the cell's interior were recognized: the disorderly, spongelike folded membrane called the

---

while the biologist Hermann Muller suggested in 1921 that the replication of genetic material in cells might parallel the templated growth of crystals. In 1943, physicist Erwin Schrödinger famously proposed that genes might be encoded in an "aperiodic crystal" – one that lacks a regular atomic organization.

* The same word is used in atomic physics to describe the dense core of the atom, where nearly all of its mass resides. I love the coincidence that Brown's observation of the dancing motions of pollen grains in water, caused by collisions with molecules, was what led ultimately to experimental proof in 1908 that atoms are real objects and not just a handy conceit for thinking about the structure of matter. Just three years after that, Ernest Rutherford discovered the atomic nucleus.

endoplasmic reticulum, and the Golgi apparatus, named after the Italian biologist Camillo Golgi. A vigorous debate began about whether the cell's jelly-like internal medium – called the proto-plasm – is basically granular, reticular (composed of membrane networks) or filamentary (full of fibrous structures). All of these options were observed, and the truth is that it depends on when and where you look at a cell.

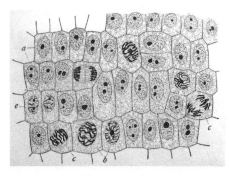

Images of onion cells drawn in 1900 by the biologist Edmund Beecher Wilson. These show some of the internal structures seen at different times in different cells. The single dense blob in several of these is the nucleus. Sometimes this seemed to break up into threads or blobs, perhaps in the process of cell division. What was going on in there?

All this internal structure led the zoologist Edmund Beecher Wilson to express regret that the term "cell" had ever been coined. It was a misnomer, he said – "for whatever the living cell is, it is not, as the word implies, a hollow chamber surrounded by solid walls." Some others wondered if the cell was really the fundamental entity it had been supposed to be, not least because a cell membrane was not always visible under the microscope. Perhaps it was actually these organized contents in the protoplasm that were the real stuff of life? "We must be careful," warned zoologist James Gray in 1931, "to avoid any tacit assumption that the cell is the natural, or even the legitimate, unit of life and function."*

At any rate, the cell contained so much stuff. What was it all for?

It was becoming clear, too, what some of the key chemical components of cells are. Chemists interested in the processes of life – by the end of the century that topic was known as *biochemistry* – had figured out that they contain chemical agents called enzymes that carry out its characteristic panoply of metabolic reactions. Some enzymes, for example, allow yeast to ferment sugar into alcohol. In 1897, German chemist Eduard Buchner showed that intact cells weren't necessary for that to happen: the "juice" that could be extracted from yeast cells could produce fermentation on its own, presumably because it still carried the crucial enzymes, undamaged, within it.

These molecular ingredients, like workers in a city, had to be organized, segregated, orchestrated in the time and place of their actions and motions. Chemical reactions in the cell have to happen in the proper order and in the right location; things can't be the same everywhere in the cell. And so the "social" view of bodies as communities of cells was repeated for the single cell itself: it was a sort of factory populated by cooperating enzymes and other molecules. This hidden machinery enables a cell to persist and maintain itself, to take in substances and energy from the environment and use them to carry out the metabolic reactions without which there is only death.†

---

* Surprisingly there are still a few advocates of that view – the "protoplasm theory" of life – today. For the fact is that the cell's wall or membrane is not merely a "housing" for all the machinery. Confinement of the contents, and their isolation from the surrounding environment, are essential for a cell to function as a living thing.

† Metabolism can, however, be put on hold in some cells and organisms in certain circumstances: they can become quiescent, in a state like suspended animation, which can be a handy survival strategy.

At the turn of the century, the substructure and organization on which the animation of cells depends was largely beyond the resolving power of the microscopes. But it was clear enough that not all cells are alike in their composition and structure. Bacteria and protists have rather little in the way of visible internal organization. They belong to a class of micro-organisms called prokaryotes, and they are typically round or elongated and sausage-shaped. The language of biological classification is always a little presumptuous, but it takes nothing away from bacteria to say that their cells are structurally relatively "simple". They lack a nucleus – hence the label "prokaryotes", meaning "pre-nucleus". (More presumption – as though bacteria just haven't yet discovered the wisdom of having a nucleus but will wake up to it one day. In fact, bacteria have existed for longer than eukaryotes; they and other prokaryotes dominate much of the planet's ecology, and evidently have no need of "greater sophistication" in order to thrive.)

Human cells, along with those of other animals, plants, fungi and yeast, are said to be eukaryotes: a term that simply connotes that their cells have a nucleus. Eukaryotic organisms may be multi-celled, like us, or single-celled, like yeast. The latter is an example of a "lower" eukaryote: more presumption, of course, but meaning that the degree of organization in the cell is less than that evident in the higher eukaryotes like peas, fruit flies and whales.

For now we can set prokaryotes aside. There is, mercifully, no need either to look in detail at what all the complex structure of the human cell is about, other than to say that it can be usefully regarded as a compartmentalization of the processes of existence. Membrane-wrapped substructures of the cell are called organelles, and each can be somewhat crudely considered to carry out a specific task. Mitochondria are the regions where a eukaryotic cell produces its energy, in the form of small molecules that release stored chemical energy when transformed by enzymes. The Golgi apparatus functions as a kind of cellular post office, processing proteins and dispatching them to where they are needed. The nucleus is where the chromosomes are kept: the material encoding the genes that are passed on when a cell divides or an organism reproduces. What we *do* need to hear more about, very shortly, are those chromosomes, because they are an important part of what defines you as an individual, and absolutely vital for orchestrating the life processes that enabled you to grow and which sustain you daily.

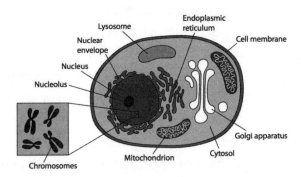

The human cell.

By the early twentieth century, it was clear that what sets living matter apart from the inanimate is not merely a question of composition: of what life is made of. Neither is it just a question of structure. Organisms and cells clearly did have a hierarchy of significant, specific yet hard-to-interpret structures reaching down to the microscopic and beyond. And that mattered. But the real reason living matter is not equivalent to some other state of matter such as liquids and gases is that it is *dynamic*: always changing, always in the process of doing something, never reaching a steady equilibrium. Staying alive is not a matter of luxuriating in the state of aliveness but is a relentless task of keeping balls in the air.

Researchers today might rightly point out that this dynamic, out-of-equilibrium character is not unique to life. Our planet's climate system is like that too: a constant channelling of the energies of the sun and of the hot planetary interior into orchestrated cycling movements of the oceans, atmosphere and sluggish rocky mantle, accompanied by flows of chemical elements and heat between the different components of the planetary system. The system is responsive and adaptive. But this is precisely the point: there *are* parallels between a living organism and the planet itself, which is why the independent scientist James Lovelock pushed the point from analogy to the verge of genuine equivalence in his Gaia hypothesis. Arguments about whether the planet can be truly considered "alive" are moot, because the living systems – rainforests, ocean microfauna, every creature that takes in chemicals and turns them into something else plus heat – are in any case a crucial, active part of the planet's "physiology".

This activity of the planetary biosphere commenced close to four billion years ago and has not ceased since. Virchow's *omnia cellula e cellula* has a significance barely any lesser than that of Darwinian evolution, which ultimately depends on it (ironically, given Virchow's views on Darwin). It establishes a basis for what Aristotle imagined as a Great Chain of Being, in which the fundamental unit is no longer the reproducing organism but the dividing cell. All cells are, in evolutionary terms, related to one another, and the question of origin reduces to that of how the first cell came into being. Since that obscure primeval event, to the best of our knowledge no new cell has appeared *de novo*.

At the same time, Virchow's slogan is a description, not an explanation. Why is a cell not content to remain as it is, happily metabolizing until its time runs out? One answer just begs the question: if that is all cells did, they would not exist, because their *de novo* formation from a chemical chaos is far too improbable. Then we risk falling back again on anthropomorphism – cells intrinsically *want* to reproduce by division – or on tautology, saying that the basic biological function of a cell is to make more cells ("the dream of every cell is to become two cells", said the Nobel laureate biologist François Jacob). Biological discourse seldom does much better than this. Cell and molecular biologists and geneticists have a phenomenal understanding of how cells propagate themselves. But explaining *why* they do so is a very subtle affair, and it's fair to say that most biologists don't even think about it. Yet that "impulse" is the engine of Darwinian evolution and consequently at the root of all that matters in biology.

There is not a goal to this process of life, towards which all the machinery of the cell somehow strives. We can't help thinking of it that way, of course, because we are natural storytellers (and because *we* do have goals, and can meaningfully ascribe them to other animals too). So we persuade ourselves that life aims to make babies, to build organisms, to evolve towards perfection (or at least self-improvement), to perpetuate genes. These are all stories, and they can be lovely as well as cognitively useful. But they do not sum up what life is about. It is a thing that, once begun, is astonishingly hard to stop; actually we do not know how that could be accomplished short of destroying the planet itself.

* * *

Life's unit is the cell. Nothing less than the complete cell has a claim to be called genuinely alive.* It's common to see our body's cells referred to as "building blocks" of tissues, much like assemblies of bricks that constitute a house. To look at the cells in a slice of plant tissue, such as Wilson's drawing of the onion earlier, you can understand why. But that image fails to convey the dynamic aspect of

_____

* A possible exception is viruses: particles still tinier than cells, made of little more than genetic material (DNA or its molecular cousin RNA) enclosed in a sheath

cells. They move, they respond to their environment, and they have life cycles: a birth and a death. They receive and process information. As Virchow suggested, cells are to some degree autonomous agents: little living entities, making their way in the world.

Anything less than a cell, then, has at best a questionable claim to be alive; from cells, you can make every organism on Earth. We have known about the fundamental status of the cell for about two centuries but have not always acknowledged it. For much of the late twentieth century, the cell was relegated before the supremacy of the gene: the biological "unit of information" inherited between generations. Now the tide has turned again. "The cell is making a particular kind of reappearance as a central actor in today's biomedical, biological, and biotechnological settings," writes sociologist of biology Hannah Landecker. "At the beginning of the 21st century, the cell has emerged as a central unit of biological thought and practice . . . the cell has deposed the gene as the candidate for the role of life itself."

Cells do more than persist. Crucially, they can replicate: produce copies of themselves. Ultimately, cell replication and proliferation drives evolution. Life is not what makes this propagation of cells possible; rather, that is what life *is*.

Biologists towards the end of the nineteenth century recognized that reproduction of cells happens not by the spontaneous formation of new cells, as Schwann believed, but by cell division as Virchow

---

of protein – as one biologist famously put it, "bad news in a protein coat". These infectious agents are able to hijack cells, forcing the cell's machinery to make copies of the virus. Whether viruses, which can't replicate and propagate on their own, can be considered truly alive is still argued over, because we lack an agreed definition of life itself. We don't need to be troubled by that debate. The simple fact is that viruses are an evolutionarily stable way for this form of organized matter to propagate itself, if cellular life is already a given. You could say that cells could in principle exist and persist without viruses, but not vice versa. Yes, you *could*; but it might not mean very much, since the existence of cells probably make viruses inevitable in the way that money makes moneylenders inevitable. Whether they (viruses, but also moneylenders) are parasites depends on where you stand.

asserted: one cell dividing in two. Single-celled organisms such as bacteria simply replicate their chromosomes and then bud in two, a process called binary fission. But in eukaryotic cells the process is considerably more complex. Cell "fission" was first seen in the 1830s and was called *mitosis* in 1882 by the German anatomist Walther Flemming, who studied the process in detail in amphibian cells.

Flemming was a champion of the filamentary model of cells – the idea that their contents are organized mainly as long fibrous structures. In the 1870s, he showed that as animal cells divide, the dense blob of the nucleus dissolves into a tangle of thread-like structures (mitosis stems from the Greek word for thread). The threads then condense into X-shaped structures that are arranged on a set of star-like protein filaments dubbed an aster. (The word means "star", but actually the appearance is more reminiscent of an aster flower.) Flemming saw that the aster gets elongated and then rearranged into two asters, on which the chromosomes break in half. As the cell body itself splits in two, these chromosomal fragments are separated into the two "daughter" cells and enclosed once again within nuclei.*

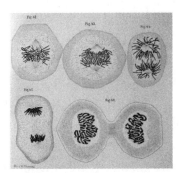

Various stages of cell division or mitosis as recorded by Walther Flemming in his 1882 book Zellsubstanz, Kern und Zelltheilung (Cell Substance, Nucleus and Cell Division).

---

* I'm putting this all a little more tidily than was apparent to Flemming.

So cell division is preceded by a reorganization of its contents: apparently, they are apportioned rather carefully into two. The thread-like material seen by Flemming unravelling from the nucleus readily takes up a staining dye (so that it is more easily seen under the microscope), leading it to be called, after the Greek word for colour, *chromatin*. The individual threads themselves were christened chromosomes – "coloured bodies" – in 1888.

In that same year, the German biologist Theodor Boveri discovered that the movement of chromosomes during cell division is controlled by a structure he called the centrosome, from which the strands of asters radiate. The two asters that appear just before a cell splits in two, each with a centrosome at their core, could in fact be seen to be connected by a bulging bridge of fine filaments, called the mitotic spindle. Flemming became convinced that these spindle fibres act as a kind of scaffold to direct the segregation of the chromosome threads into two groups. He was right, but he lacked a sufficiently sharply resolved microscopic technique to prove it.

So the division of animal cells isn't just like the splitting of a water droplet into two. It has to be accompanied by a great deal of internal reorganization. Flemming and others identified a series of distinct stages along the way. While cells are going about their business with no sign of dividing, they are said to be in the interphase state. The unpacking of the nucleus into filamentary chromosomes is called prophase, and the formation and elongation of the aster is called metaphase. As the aster-like cluster splits in two, the cell enters the anaphase, from where it is downhill all the way to fission and the re-compaction of the nucleus.

This procedure is called the *cell cycle*, which is an interesting phrase when you think about it. Its implication is that, rather than thinking of biology as being composed of cells that do their thing until they eventually divide, we might regard it as a process of continual replication and proliferation that involves cells. With all due warning about the artificiality of narratives in biology, we might thus reframe the Great Chain of Being as instead a Great Chain of Becoming.

*   *   *

It was a fundamental – perhaps *the* fundamental – turning point for modern biology when, around the turn of the century, scientists came to appreciate that much of the complicated reorganization that goes on when cells divide is in order to pass on the genes, the basic units of inheritance, that are written into the strands called chromosomes. What they were seeing here in their microscopes is the underlying principle that enables inheritance and evolution.

The notion of the gene as a physical entity that confers inheritance of traits appeared in parallel with the development of cell theory in the mid-nineteenth century. The story of how "particulate factors" governing inheritance were posited by the Moravian monk Gregor Mendel from his studies on the cultivation of pea plants has been so often told that we needn't dwell on it. In the 1850s and '60s Mendel observed that inheritance seemed to be an all-or-nothing affair: peas made by interbreeding plants that make smooth or wrinkly versions are either one type or the other, not a blend ("a bit wrinkly") of the two. Of course, real inheritance in humans is more complicated: some traits (like hair or eye colour) may be inherited discretely, like Mendel's peas, others (like height or skin pigmentation) may be intermediate between those of the biological parents. The puzzle Mendel's observations raised was why inheritance is not always such mix, given that it comes from a merging of the parental gametes.

Charles Darwin didn't know of Mendel's work, but he invoked a similar idea of particulate inheritance in his theory of evolution by natural selection. Darwin believed that the body's cells produced particles that he called gemmules, which influence an organism's development and are passed on to offspring. In this view, all the cells and tissues of the body play a role in inheritance, whence the term "pangenesis" that Darwin coined for his speculative mechanism of evolution. These gemmules may be modified at random by influences from the environment, and the variations are acquired by progeny. In the 1890s, the Dutch botanist Hugo de Vries and German biologist August Weismann independently modified Darwin's theory by proposing that transmission of gemmules could not occur between body (somatic) cells and the so-called "germ cells" that produce gametes. Only the latter could contribute to inheritance. De Vries used the term "pangene" instead of gemmule to distinguish his theory from Darwin's.

At the start of the twentieth century, the Danish botanist Wilhelm Johannsen shortened the word for these particulate units of inheritance to "gene". He also drew the central distinction between an organism's *genotype* – the genes it inherits from the biological parents – and its *phenotype*, the expression of those genes in appearance and behaviour.

In 1902 Theodor Boveri, working on sea urchins in Germany, and independently the American zoologist Walter Sutton, who was studying grasshoppers, noticed that the faithful passing on of chromosomes across generations of cells mirrored the way that genes were inherited. Perhaps, they concluded, chromosomes are in fact the carriers of the genes. Around 1915, the American biologist Thomas Hunt Morgan established, from painstaking studies of the inheritance of characteristics in fruit flies, that this is so. Moreover, Morgan showed how one could deduce the approximate positions of two different genes relative to one another on the chromosomes by observing how often the two genes – or rather, the manifestation of the corresponding phenotypes – appear together in fruit flies made by mating of individuals with the respective genes. As the chromosomes were divvied up to form egg and sperm cells, genes that sat close together were more likely to remain together in the offspring. Morgan's work established the idea of a genetic map: literally a picture of where genes sit on the various chromosomes.

The sum total of an organism's genetic material is called its *genome*, a word introduced in 1920. For many years after Morgan's work, it was suspected that genes are composed of the molecules called proteins, in which the much smaller molecules called amino acids are linked together in chains. Proteins, after all, seemed to be responsible for most of what goes on in cells – they are the stuff of enzymes. And chromosomes were indeed found to consist partly of protein. But those threads of heredity were also known to contain a molecule called DNA, belonging to the class known as nucleic acids (that's what the "NA" stands for).

No one knew what this stuff did until the mid-1940s, when the Canadian-American physician Oswald Avery and his co-workers at the Rockefeller University Hospital in New York reported rather conclusive evidence that genes in fact reside on DNA. That idea was not universally accepted, however, until James Watson, Francis Crick, Maurice Wilkins, Rosalind Franklin and their co-workers revealed the molecular structure of DNA – how its atoms are arranged along the chain-like molecule. This structure, first reported in 1953 by Watson and Crick, who relied partly on Franklin's studies of DNA crystals, showed how genetic information could be encoded in the DNA molecule. It is a deeply elegant structure, composed of two chain-strands entwined in a double helix.

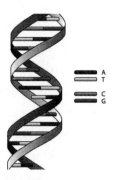

The double helix of DNA. This iconic image creates a somewhat misleading picture, since for most of the time DNA in a cell's chromosomes is packaged up quite densely in chromatin, in which it is wrapped around proteins called histones like thread on a bobbin. The "rungs" of the double-helical ladder consist of pairs of so-called nucleotide bases (denoted A, T, C and G) with shapes that complement each other and fit together well.

So beautiful, indeed, was this molecular architecture and the story it seemed to disclose that modern biology was largely seduced by it. It was immediately obvious to Watson and Crick how heredity could be enacted on the molecular scale. The information in genes could be replicated by unzipping the double helix so that each strand could act as the template on which replicas could be assembled.* Here, then, was how genetic information could be copied into new chromosomes when cells divide: a molecular-scale mechanism for the inheritance described by Mendel and Darwin,

---

* Not exact copies of each strand, however, but "complementary strands". Each molecular building block in one strand is paired with a partner in the other, like men and women matching up in pairs in a quadrille. So during replication of chromosomes, a single DNA strand acts as the template for formation of the complementary strand by mutual recognition of these pairs of building blocks. In this way, the twin strands of a double helix together assemble the strands of another double helix with the same sequence of building blocks.

which Morgan and others had situated on the chromosomes. DNA married genetics with inheritance at the molecular level, bringing coherence to biology.

And Darwinian evolution? If genes govern an organism's traits, then random copying errors in DNA replication could alter a trait, mostly to the detriment of an organism but occasionally to its advantage. This is the variation on which natural selection acts to make organisms adapted to their environment.

It all seemed to fall into place. All the important questions – about evolution, genetic disease, development – might now be answered by referring to the information in the genome. Cells didn't seem to be a very important part of the story except as vehicles for genes and as machines for enacting their commands.

To speak of information being "encoded" in DNA is to speak literally. Genes deploy a code: the genetic code. But what exactly do genes encode? On the most part, it is the chemical structure of a protein molecule, typically an enzyme. Because of the ways in which different amino acids "feel" one another and interact with the watery solvent all around them in the cell, a particular sequence of amino acids determines the way most protein chains fold up into a compact three-dimensional shape. This shape enables enzymes to carry out particular chemical transformations in the cell: they are catalysts that facilitate the cell's chemistry. So the protein's sequence, encoded in the respective gene, dictates its function.

A protein's amino-acid sequence is represented in its gene by the sequence of chemical constituents that make up DNA. There are four of these, called nucleotide bases and denoted by the labels A, T, G and C. Different triplets of bases represent particular amino acids in the resultant protein: AAA, for example, corresponds to the amino acid called lysine.

Turning a gene into its corresponding protein is a two step-process. First, the gene on a piece of DNA in a chromosome is used as a template for building a molecule of another kind of nucleic acid, called RNA. This is called transcription. The piece of RNA made from a gene is then used as a template for putting the protein together, one amino acid at a time. This is called translation, and it is performed by a complex piece of molecular machinery called the ribosome, made of proteins and other pieces of RNA.

Chromosomes consist of lengths of DNA double-helix wound around disk-like protein molecules called histones, like the string on a yoyo. This combination of DNA and its protein packaging is what we call chromatin. The genomes of eukaryotes are divided up into a number of chromosomes that is always the same for every cell of a particular species (if they are not abnormal) but can differ between species. Human cells have 46 chromosomes, in sets of 23 pairs.

* * *

It's common to see genes called the instructions to make an organism. In this view, the entire genome is then the "instruction booklet", or even the "blueprint". This is an understandable metaphor, but misleading. Genes are fundamental to the way an organism turns out: the genome of a frog egg guides it to become a frog, not an elephant, and vice versa. But the way genes influence and to some degree dictate that proliferation of cells is subtle, complex, and resistant to any convenient metaphors from the technological world of design and construction. By leaping from genome to finished organism without taking into account the process of development from cells, we risk simplifying biology in ways that can create some deep misconceptions about how life proceeds and evolves.

To the extent that a gene is an "instruction", it is an instruction to build a protein molecule. It is far from obvious what, in general, this has to do with the growth and form of an organism: with the generation of our flesh. We know of no way to map an organism's complement of proteins onto its shape, traits and behaviour: its phenotype. The two are worlds apart: it's rather like trying to understand the meaning of a Dickens novel from a close consideration of the shapes of its letters and the correlations in their order of appearance.

Besides, this conventional "blueprint" description of what genomes do is too simplistic even if we consider only how they dictate that roster of proteins. Here are some reasons why:

- Only about 1.5 per cent of the human genome encodes proteins, and a further 8 to 15 per cent or so is thought to "regulate" the activity of other genes by encoding RNA that turns their transcription up or down. We don't know what the rest does, and scientists aren't agreed on whether it is just useless "junk" accumulated, like rubbish in the attic, over the course of evolution, or whether it has some unknown but important biological function. In all probability, it is a bit of both. But at any rate, a lot of this DNA with no known protein-coding or regulatory function is nonetheless transcribed by cells to RNA, and no one is sure why.

- Most protein-coding human genes each encode more than one protein. Genes are not generally simply a linear encoding of protein sequences that start at one end of the protein chain and finish at the other; they are, for example, interspersed with sequences called introns that are carefully snipped out of the transcribed RNA before it is translated. Sometimes the transcribed RNA then gets reshuffled before translation, providing templates for several different proteins.

- Proteins are not just folded chains of amino acids. Sometimes those folded chains are "stapled" in place by chemical bonds, or clipped together by other chemical entities such as electrically charged ions. Most proteins have other chemical groups added to them (by other enzymes) – for example, a group containing an iron atom is needed by the protein haemoglobin to bind oxygen and carry it around the body in the blood. None of these details, essential to the protein's structure and function, is encoded in DNA. You would not be able to deduce them from a gene sequence.

- We only know what around 50 per cent of gene-encoded proteins do, or even what they look like. The rest are sometimes called "dark" proteins: we assume they have a role but we don't know what it is.

- Plenty of proteins do not seem to have well-defined folded states but appear loose and floppy. Understanding how such ill-defined "intrinsically disordered proteins" can have specific biological roles is a very active area of current research. Some researchers think that the floppiness may not reflect the state of these proteins in cells themselves – but we don't really know if that is so or not.

Ah, details, details! How much should we care? Do they really alter the picture of genes dictating the organism?

That depends, to some degree, on what questions you are asking. A genome sequence – the ordered list of nucleotides A, T, C and G along the DNA strands of chromosomes – does specify the nature of the organism in question. From this sequence you can tell in principle if the cell that contains it is from a human, a dog or a mouse (something that may not be obvious from a cursory look at the cell as a whole). These distinctions are found only in some key genes: the human genome differs from that of chimpanzees in just 1 per cent of the sequence, and a third of it is essentially the same as the genome of a mushroom.* The differences between the genomes of individual people are even tinier.

But whereas you can look at a real blueprint, and probably an instruction manual, and figure out what kind of object will emerge from the plan, you can't do that for a genome. Indeed, you can

---

* The common quip that we are therefore "one-third mushroom", or "three-fifths banana" or whatever, is a rather good illustration of the foolishness of equating

only deduce that a genome will "produce" a dog at all if you have already decoded the generic dog genome for comparison, laying the two side by side. It's simply a case of seeing if the two genomes superimpose; there's nothing intrinsic in the sequence that hints at its "dogness".

This isn't because we don't yet know enough about the "instructions" in a genome (although that is the case too). It's because there is no direct relationship between the informational content of a gene – which, as I say, typically dictates the structure of a class of protein molecules, or at least of the basic amino-acid fabric of those proteins – and a trait or structure apparent in the organism. Most proteins do jobs that can't easily be related to any particular trait. Some can be: for example, there's a protein that helps chloride ions get through the membranes of our cells, and if this protein is faulty – because of a mutation in the corresponding gene – then the lack of chloride transport into cells causes the disease cystic fibrosis. But in general, proteins carry out "low-level" biochemical functions that might be involved in a whole host of traits, and which might have very different outcomes if the protein is produced ("expressed") at different stages in the development or life cycle of the organism. As microbiologist Franklin Harold has said, "the higher levels of order, form and function are not spelled out in the genome."

Might we, then, call a genome not a blueprint but a recipe? The metaphor has rather more appeal, not least because many recipes assume implicit knowledge (especially in older cookbooks). But a recipe is still a list of ingredients plus instructions to assemble them. Genomes do not come with users' instructions, more's the pity.

---

genomes with organisms, ignoring the unifying principle of the cell. What these equations really mean is that all cells have certain fundamental metabolic requirements, but that slight adjustments to their developmental programmes are enough to guide them into assembling in strikingly different ways: into organisms with very different shapes, behaviours and capabilities.

Harold offers a different image, allusive and poetic and all the more appealing for that:

> I prefer to think of the genome as akin to Hermann Hesse's *Magister Ludi* [aka *The Glass Bead Game*]: master of an intricate game of cues and responses, in which he is fully enmeshed and absorbed; a game that is shaped as much by its own internal rules as by the will of that masterful player.

If there was better public communication of the complex, contingent and often opaque relationship of genotype to phenotype, there might be rather less anxiety about the idea that genes affect behaviour. Small variations in each individual's genetic make-up can have an influence – sometimes a rather strong one – not just what you look like but what your behaviour and personality are like. This much is absolutely clear: there is not a single known aspect of human behaviour so far investigated that does not turn out to show some correlation with what gene variants we have. Even habits or experiences as apparently contingent and environmental as the amount we watch television* or our chance of getting divorced are partly heritable, meaning that the differences between individuals can be partly traced to differences in their genes.

Far from alarming us, this shouldn't surprise us. We have always been content to believe that, for example, some people seem blessed with talents that can't obviously be explained by their environment and upbringing alone. By the same token, some seem

---

* This sounds absurd, because television is far too recent an invention for our watching habits to have resulted in any evolutionary selection among genes. But that's not the point. Our tendency to watch television will depend on far more general cognitive traits, such as attentiveness and curiosity, that we have every reason to believe have some very ancient adaptive significance.

hardwired to find particular tasks challenging, such as reading or spatial coordination.

Yet perhaps because we have a strong sense of personal agency, autonomy and free will, many people are disturbed by the idea that there are molecules in our cells that are pulling our strings. They needn't worry. It is precisely because genetic propensities are filtered, interpreted and modified by the process of growing a human cell by cell that they don't fully determine how our bodies turn out, let alone how our brains get wired . . . let alone how we actually behave.

Genes supply the raw material for developing our basic cognitive capabilities – to put it crudely, they are a key part of what allows most human embryos to grow into bodies that can see, hear, taste, that have minds and inclinations. But how they exert their effects is very, very complicated. In particular, very few genes affect one trait alone. Most genes have influences on many traits. Some traits, both behavioural and medical (such as susceptibility to heart disease), seem to be influenced – in ways that are imperceptible gene by gene, but detectable when their effects are added up – by most of the genome. That's why the popular notion of a "gene for" some behavioural trait is misguided. In fact, it means that there may be no meaningful "causal" narrative that can take us from particular genes to behaviours.

* * *

This is precisely why we need to resist seductively simple metaphors in genetics: blueprints, selfish genes, "genes for". Of course, science always needs to reduce complex ideas and processes to simpler narratives if it is going to communicate to a broader audience. But I've yet to see a metaphor in genomics that does not risk distorting or misrepresenting the truth, so far as we currently know it. Fortunately, I do not think this matters for talking about the roles of genes in making a human. We will deal with those roles as they arise, without resorting to any overarching story about what genes "do".

I haven't even told you yet the worst of it, though. It's not simply difficult to articulate clearly what, in the scheme of growing humans the natural way, genes do. For *we don't exactly know how to define a gene at all.*

This isn't a failure of biology, but a strength. It's tempting to imagine that science can't be fully coherent if it can't define its key terms. But the most fundamental concepts are in fact almost invariably a little hazy. Physicists can't say too precisely or completely what time, space, mass and energy are. Biologists can't say what a gene or a species is. For that matter, chemists aren't fully agreed on what an element or a chemical bond is. In all cases, these terms arose because it seemed as though they had a very specific meaning, but when we looked more closely we found fuzzy edges. Yet the reason we coined the terms in the first place was because *they were good for thinking with.*

That remains true. A gene is a useful idea, perhaps in much the same way as words like "family" and "love" and "democracy" are useful: they are vessels for ideas that enable us to have useful conversations. They are usually *precise enough.*

Here, then, is a definition good enough to let us talk about genes in the role of growing a human from cells. Think of a gene as a piece of DNA from which a cell is able to make a particular molecule, or group of molecules, that it needs in order to function. By passing on copies of genes, cells can pass on that information so that the progeny doesn't have to rediscover it from scratch.

If you raised your eyebrows at "so that", good for you. In such phrases, biology is given a false purpose, an illusory sense that it pursues goals. It is nigh on impossible to talk about biology – about growth, development, evolution – without some mention of aims. Try to remember that this is always mere metaphor. The way that the laws of the physical world have played out on our planet is such that entities called cells have appeared that have a propensity to pass on genes to copies of themselves. This is remarkable and marvellous. No one really understands why it happens – why reproduction, inheritance and evolution is possible – and that's why we find it necessary to tell stories about it. All we can say is that there is absolutely nothing that forces us to invoke any supernatural explanation for it. The gaps that remain would make an extremely peculiar shape into which such an account might be squeezed.

Here's another thing worth knowing about genes: *a gene on its own is useless*. It can't replicate,* it can't even do the job that evolution "appears" to have given it. Frankly, there is no real point in calling a gene on its own a "gene" at all: the name connotes an ability to (re)produce, but a lone "gene" is sterile, just a molecule that happens to resemble a part of the DNA in a chromosome. It's common to say that a gene is a piece of DNA with a particular sequence, but the truth is that such a physical entity only becomes a gene in the context of a living system: a cell, at minimum. Genes are central ingredients of life, but by the time you reach the level of the gene there is nothing left that is meaningfully alive.

No, life starts with the cell. And that's why a gene only has meaning by virtue of its situation in a cell. Does this, then, mean that the cell is more fundamental to biology than the gene? You might as well ask if words are more fundamental to literature than

---

* One can sometimes read, in books that are otherwise excellent on the topic of genes and evolution, the assertion that genes are replicators. This is wrong, by any meaningful definition of the word. No gene has ever been shown to be able to replicate autonomously if given the component molecules from which it may be assembled. Aside from human technologies for doing the job, genes get replicated only by enzymes working within the context of cells. Genes are not replicators but *replicands*: that which is replicated in the course of life.

stories. It is "stories" that supply the context through which words acquire meaning, making them more than random sounds or marks on paper.

And by "context" here I don't just mean that a gene has to be in a cell in order to represent any biologically meaningful information. I mean also that, for example, the history of the cell, and of the entire organism, might matter to the function a gene has. A gene that is "active" at one point in the organism's growth might represent a quite different message – have a different implication – than at a later or earlier point. Yet the molecular machine (the protein) encoded by the gene may be identical in the two cases. The gene doesn't change, but the "instruction" it represents does.

You might compare it to the exclamation "Stop it!" Is that an instruction? Well, of course you don't know from that enunciation alone what it is you are supposed to stop, but perhaps you might regard it as a generic instruction to desist from the activity you're engaged in. But what if you hear someone shout "Stop it!" as you see a football rolling towards a cliff edge? Is that an instruction to desist in anything, or on the contrary an injunction to action? You need to know the context.

*　*　*

The gene-centred narrative of life is just one example of our urge to somehow capture the essence of this complicated, astounding process – to be able to say "life starts here!" Science's reductionist impulse gets a bad press, but breaking complicated things down into simpler ones is a tremendously powerful way of making sense of them. I think that what many people who complain about reductionism are reacting against is not so much this process of analysis – of taking apart – in itself, but the tendency then to assert "this is what *really* matters". Science has sometimes been a little slow to recognize the problem with such assertions. When one group of physicists started insisting it was going to find a Theory of Everything – a set of fundamental laws from which the entire physical universe emerged – others pointed out that it would be nothing of the kind, because it would be useless in itself for predicting or explaining most of what we see in the world.

It's not just that we should resist the temptation to see reductive analysis as a quest to identify what is more important/fundamental/real in the world. Sometimes the phenomenon you're interested in *only exists at a particular level in the hierarchy of scale*, and is invisible above and below it. Go to quarks and you have lost chemistry. With genes and life it is not quite that extreme – but at the level of genes you are left with only a rather narrow view of some of the entities and processes that underpin this notion we call "life". Life remains a meaningful idea from the macro level of the entire biosphere of our planet right down to the micro level of the cell. Within those bounds it encompasses a whole slew of factors: flows of energy and materials, the appearance of order and self-organization, heredity and reproduction. But below the level of the cell, you'll always be overlooking something vitally important in life. As Franklin Harold has put it:

> Something is not accounted for very clearly in the single-minded dissection to the molecular level. Even as the tide of information surges relentlessly beyond anyone's comprehension, the organism as a whole has been shattered into bits and bytes. Between the thriving catalog of molecules and genes, and the growing cells under my microscope, there yawns a gulf that will not be automatically bridged when the missing facts have all been supplied. No, whole-genome sequencing won't do it, for the living cells quite fail to declare themselves from those genomes that are already in our databases . . . The time has come to put the cell together again, form and function and history and all.

It is precisely the multivalent, multiscale implication of the word "life", too, that creates the tensions, ambiguities and ambivalences about what it means to "grow a human". We are thereby "making a life", but not "making life". That same truth is spoken in jest in a cartoon by Gary Markstein in which two white-coated scientists contemplate IVF embryos. "Life begins at the Petri dish!" exclaims one embryo; "Cloning for research!" demands another. "Even the human embryos are divided", sighs a scientist.

This is the struggle we face in reconciling our notion of life as human experience with the concept of life as a property of our material substance. We are alive, and so is our flesh. While those two visions of life were synonymous, we could ignore the problem. Having a mini-brain grown in a dish from a piece of one's arm tends to make that evasion no longer tenable.

It's no wonder that different cultures at different times have had such diverse attitudes to the connection between the human body *in utero*, forming in hidden and mysterious fashion from something not remotely human-like, and the human body in the world. The insistence by some people and in some belief systems that "life begins at conception" is a modern utterance, often claiming firm support from the very science that in fact shows how ill-defined the idea is.

But the tension is an old one, as demonstrated by preformation theories of the human fetus. This was an anthropomorphization of the cell as explicit as that in cartoons that attribute voices and opinions to human embryos in petri dishes. Intuition compels us to look for the self in the cell. An insistence on locating it instead in our genes – as cell biologist Scott Gilbert puts it, to see "DNA as our soul" – comes from the same impulse. Perhaps we must be gentle in dispensing with these superstitions. Aren't old habits always hard to shake off?

CHAPTER 2

BODY BUILDING

GROWING HUMANS THE OLD-FASHIONED WAY

So far, nothing beats sex. Biologically, I mean. If you want to grow a human, you need a sperm and an egg cell – the two cell types called gametes. And you need to get them together. That's an objective towards which an immense amount of our culture is geared.

In describing the process in which a fertilized egg develops into a person, I hope in this chapter to give back some of the strangeness, the proper unfamiliarity, to embryology: to show how removed our individual origins are from the comforting intimacy of the gracefully curled fetus that is generally our first ultrasonic glimpse of a new human person.

We are folded and fashioned from flesh in its most basic form, according to a set of instructions that is far removed from a kind of genetic step-by-step. We are shaped from living clay according to rules imperfectly known and often imperfectly executed, and which orchestrate a dance between the cell and its environment.

But there are many things you can potentially fashion from clay, if you know how to work the wheel. As we come to understand more about the emergence of the human *ex ovo*, we perceive new possibilities, new beginnings and routes and directions. And we change from watchers to makers.

* * *

There is no new narrative of human-growing that does not need to reckon with the preconceptions (so to speak) created by sex. Mary Shelley could not, in her day, make that context explicit – but Victor Frankenstein's terror of his wedding night tells us all about the psychosexual undercurrents in his onanistic act of creation. I won't therefore attempt the same evasion as the school biology lesson by beginning the embryo's tale with sperm meeting egg; by that stage sex has, as we'll see, already imposed itself on the story.

We should in any case be continually amazed, surprised and possibly even a little proud at how imaginatively we have elaborated, ritualized and celebrated the urge to procreate. This shouldn't be seen so much as proof that evolutionary psychology can "explain" culture – the banal observation that because of our instinct for sexual reproduction we write stories like *Romeo and Juliet* and create entertainments like *Love Island* – but rather the opposite. Evolutionary psychology by itself offers a rather threadbare and reductionist narrative for understanding the rich tapestry of culture. Sure, we can attribute to the sexual drive everything from a worship of *lingams* in Indian tradition to the Tudor enthusiasm for prominent red codpieces,* the hegemony and variety of internet porn and the exquisite faux-pheromone concoctions of perfumeries. But then we will have not really said much that illuminates the particulars of any of those diverting cultural phenomena, will we?

---

* When I was once preparing an exhibition trail at the V&A Museum in London that explored the "science" of some objects on display, the fabrics department curator was horrified at the suggestion that codpieces had anything to do with sexual display. The very idea!

It's tempting to suppose that the bare biology of reproduction is quite distinct from the human mechanics and its attendant rituals, its messiness and epiphanies and calamities. But we rarely make any statement about biology, and least of all about the biology of making humans, that is devoid of a culturally shaped narrative. If we imagine we can start talking about new ways to grow humans (and parts of humans) that do not inherit some aspects of the stories we tell about how we do it already, we are fooling ourselves.

The old ideas of generative male seed quickening the passive female "soil" are evidently invested with patriarchal stereotypes. Within Christian tradition, conception long struggled to find accommodation with religious thought, being simultaneously a miraculous gift of God (and thus a moral obligation) and the fruit of sin. Within this view, the only "pure" conception in the history of humankind was one that took place two thousand years ago without intercourse and without seed, to dwell on the gestation of which was to risk heresy. And medieval theology willingly lent authority to the idea that to expel male seed not directed towards procreative possibility was even worse than to couple in lust, for it was liable to be taken up by demonic succubi and bred into monsters.

These were tales not just about the social side of sex but about its biological and medical aspects too. Until the nineteenth century, the health hazards of masturbation were considered a plain medical fact, as was the idea that a fetus in the uterus could be damaged by a mother's bad thoughts. Probably every age has imagined itself mature beyond this mixing of science with folk belief and sociopolitical ideology, but let's not make the same mistake.

So how *does* the sperm fertilize the egg? Why, the plucky little fellow has to race along the vaginal passage,* out-swimming its

---

* Well, we can try starting the tale from here, but children aren't so easily satisfied. It was no good my telling my three-year-old daughter that the sperm gets

(his, surely!) peers in a Darwinian competition for survival. There sits the egg, plump and alluring – and in he dives, kicking off the process of becoming one of us. As *Life*'s science editor Albert Rosenfeld wrote in 1969, people are made from "the sperm fresh-sprung from the father's loins, the egg snug in its warm, secret place; the propelling force being conjugal love." (I think you'll find it is actually hydrogen ions crossing cell membranes.)

We see this story not just in children's books about how babies are made, but (less obviously) in some biology textbooks too, where the active role of the sperm and the passivity of the egg cell is typically stressed. It's wrong. There is now good reason to think, for example, that the sperm's entry into the egg is actively mediated by the egg (although even *this* description still somewhat anthro-pomorphizes the participants, imputing aims and roles). The fastest sperm are not necessarily the victors, because sperm needs condi-tioning by the female reproductive tract to make it competent to fertilize an egg. There is increasing evidence that in many species the female can influence which sperm is involved in fertilization – for example by storing sperm from several mates under selective conditions, or ejecting it after sex. Some experiments on genetic outcomes of fertilization seem to imply the egg somehow selects sperm with a particular genotype, defying the conventional idea that once it gets to this stage the union of gametes is random.

It's by means of narratives about sperm and egg getting together that we instruct our children about the "facts of life" – the definitive phrase seemingly designed to fend off the awkward questions they might otherwise ask. Questions like: Why *this* way? For doesn't it seem an awfully messy, contingent and chancy way for genes to propagate, requiring a costly investment in wardrobe, grooming products and

---

into the vagina by coming out of the penis. "No it doesn't," she protested. "They're too far apart."

expensive meals? If children knew that parthenogenesis (development of the ovum without fertilization) was a reproductive option in the animal kingdom, I suspect many would think it dreadfully unfair that it is not available to humans. (Not only children, actually.)

So why *these* facts? Why go through the rigmarole of sex, if it doesn't seem to be a biological *sine qua non* of reproduction?

No one really knows the answer. The usual one is that sexual reproduction, by combining the genes of two individuals, permits a beneficial reshuffling that can help stave off genetic disease. Organisms like bacteria that reproduce by simple cell division, generating clones, will gradually accumulate gene mutations from one generation to the next. As most mutations are detrimental or at best have no discernible effect on fitness, this can't be a good thing. But bacteria proliferate exponentially and rapidly, and it doesn't much matter if many genetically disadvantaged lineages die off, so long as there are a few that acquire mutations that improve fitness. In a rapidly growing bacterial colony, the mutations give the population a chance to explore a significant amount of "gene space" and find good adaptations. It's for much the same reason that bacteria have evolved mechanisms to transfer genes directly from one to the other in a non-hereditary process called horizontal gene transfer.

Organisms that reproduce slowly and sparsely, like humans, don't enjoy a bacterium's capacity to "try out" many mutations. But sex is a way to spread them more rapidly, by allowing new combinations of gene variants to be created at a stroke from one generation to the next.

Clonal reproduction is also risky as it tends to put all a population's eggs in one basket (to rather abuse a metaphor). Along comes some virus that exploits a bacterium's vulnerabilities, or a change in conditions such as drought or extreme cold, and the whole colony could be wiped out – unless a few individuals are fortunate enough to possess gene variants that can withstand the threat. That's also why genetic diversity is good for a population. And again, sexual recombination of genomes provides some of that.

So sex is a way for slowly reproducing organisms like us to eject "bad" genes and acquire "good" ones. It recommends that the organisms become dimorphic – that there be two distinct sexes, to ensure that an organism doesn't end up combining its *own* sex cells, which would rather defeat the object. Or at any rate, there must be more than one sex: there's no obvious reason why it should be limited to two, and indeed some fungi have thousands of different "sexes".*

From this basic requirement for a distinction between types of sex cell (gametes) stems all the rest of the exciting and confusing features of sexual dimorphism. It helps if the two sexes are distinguishable at a glance, to save the wasted effort (since generally it does cost time and effort, sometimes considerably so) of trying to mate with another individual with which that is not possible. (It also makes complete sense that those impulses and signals will not be equally strong, or present at all, in all individuals, making homosexuality a natural and common phenomenon in animals.)

Once those differences exist, they are apt to get amplified, diversify, sometimes way out of proportion. If you're going to have sex, it's likely that your behavioural traits will evolve to let you spot the fittest partners and to advertise your own fitness. Each sex will evolve methods of assessing mates, and outward indicators of fitness will elicit attraction in the opposite sex. Some of these make physiological sense: a lot of muscle suggests dominant males with good survival skills, wide hips in females imply superior child-bearing capacity.

* Sexual dimorphism doesn't necessarily require rigid division of the species into males and females. Plenty of animals, such as molluscs, are hermaphrodite, possessing both sets of reproductive organs. Some fish, such as clownfish, can change sex. There is typically one dominant female in a group, larger than the males. If she is removed – eaten, say, by a predator – then a male will develop ovaries and take her place, growing accordingly in stature.

Other sexual signals may end up being rather arbitrary – it's not obvious why body hair on our male ancestors would of itself confer any survival advantage. (Perhaps this was an example of "useless" signalling of fitness, like the peacock's tail?) Some displays are simply about standing out from the crowd, like exotic bird plumage. Other facets of sexual attractiveness may be subtle: it seems likely, for example, that symmetrical facial features indicate that one's developmental processes, which commonly unfold independently in the mirror-image halves of the body, are robust against random variations, giving the individual better health prospects. For all the elaborateness of some mating rituals in other animals, they might – if they could – count themselves lucky that these sexual signals and responses don't get refracted through culture as they do with humans to the point where it can all get overwhelmingly confusing.

Now, this is certainly one way to talk about the evolution and origin of sex, but it invokes an uncomfortable amount of teleology. Sex doesn't really exist *in order to* create genetic diversity. Nothing happens in evolution *in order to* produce a particular end result. It makes intuitive sense for us to speak like this, but the fact is that sex evolved because those early organisms that became able to fuse their cells and chromosomes somehow produced more robust populations than those that lacked this ability. Sexual reproduction *might* be a more or less inevitable consequence of evolution by natural selection, once it gives rise to a particular kind of complex organism, much as snowflakes are an inevitable consquence of the laws of physics and chemistry playing out in a particular environment. Evolutionary biologists say that sex is a successful *evolutionary strategy*, although this again imputes a sort of foresight to evolution that it doesn't possess.

While these arguments for the value of sex surely have a lot going for them, they can't be the whole answer. Sex is not essential for all higher vertebrates. Parthenogenesis occurs in many different types of animal, including insects such as mites, bees and wasps, and some fish, reptiles and amphibians. In a few such cases, reproduction can happen either with sex or without. Sometimes that's by design – for example, parthenogenesis is thought to be an option for mayflies as a defence against a lack of males. (The same useful trait arises in the women of the male-free society in Charlotte Perkins Gilman's utopian feminist novel *Herland* (1915).) In other cases it occurs by accident, unfertilized eggs just happening rarely to develop into embryos. Komodo dragons are among the larger creatures that can reproduce this way.

The reasons and mechanisms for parthenogenesis are varied and sometimes rather complicated. But one thing you can say for sure is that, in organisms for which it can take place, evolution has not seen fit to rule it out. To put it another way, there is nothing obviously necessary about sex, and assessing the benefits of sexual reproduction over other methods of propagating is a subtle and perhaps context-dependent business.* As far as evolution is concerned, it is just a matter of whatever works.

* * *

There's a complication with sex. Each of our body cells has two sets of chromosomes, and therefore dual copies of each gene, one inherited from each parent. But if one of these cells in a female simply fused with one from a male, the resulting cell would have four sets of chromosomes. That's too many, and the cell couldn't function properly. So organisms that reproduce sexually have evolved special kinds of cells that possess only one copy of each chromosome. These are the gametes, and they are found only in the gonads: the ovaries and testes.

---

* It's possible, though, that occasional sex *is* necessary for the long-term viability of eukaryotes. We don't know for sure.

Gametes are made from a specialized type of cell called a germ cell. The germ cells have doubled chromosomes (they are said to be *diploid*) just like somatic cells, but in a special kind of cell division called meiosis they divide into gametes in which these chromosomes have been carefully segregated into two. Cells with just a single set of chromosomes are said to be *haploid*.

Normal cell division (mitosis) involves replication of the chromosomes accompanied by their separation so that each daughter cell receives the full complement. Meiosis is even more complicated, because the existing chromosomes have to be divided precisely in two and shipped off to their respective destinations.

Actually it's worse than that. Meiosis happens in two stages, and the overall result is that a single, diploid germ cell replicates its chromosomes once and divides twice to end up with four haploid gametes. As in mitosis, the process by which the chromosomes are divided uses a spindle-like structure made from fibrous protein. The chromosomes become attached to the fibres and are drawn towards opposing poles of the spindle located in the two lobes of the dividing cell.

Crucially, the chromosomes undergo some shuffling in this process. The pre-meiosis germ cell, recall, has one of each of the 23 types of chromosome from the mother, and a second copy of each from the father. Which of the poles of the spindle each chromosome is drawn towards is random, and so the diploid cells made by division of the germ cell have a random combination of maternal and paternal genes.* The haploid gametes that eventually emerge from the process then have a thoroughly scrambled single set of chromosomes: with 23 pairs of chromosomes in all, there are 2^{23}, or about 8 million, possibilities. These are combined with a similar range of

---

* The maternal and paternal chromosomes can also exchange genes during meiosis, adding to the permutations that may result.

options in the other gamete when egg and sperm unite, so you can see that having sex is a good way to produce genetic diversity.

Formation of the so-called primordial germ cells happens early in the development of a human embryo, around two weeks after fertilization. This is even before the gonads have started to form, which is to say, before the embryo has yet "woken up" to which sex it is. It's as if the embryo is putting these cells aside while deferring the matter of whether they will be eggs or sperm. The gonads themselves will guide this process, sending out chemical signals that tell the primordial germ cells which sort of gamete to become. They're ready to do that around week six of gestation, by which time the germ cells have migrated across the developing embryo to their destination. For yes, that development involves not merely cell division but also cell movement, a physical sorting in space to arrange the parts in the proper disposition.

Germ cells were first postulated by the German zoologist August Weismann in his 1892 book *The Germ-Plasm: A Theory of Heredity*. As that title suggests, this was a hypothesis as much about evolution as about embryology. The "plasm" here reflects the widespread notion, before Boveri and Sutton's chromosomal theory of inheritance, that heredity was somehow transmitted via the "protoplasm" substance inside cells. As we saw earlier, Charles Darwin speculated that the particles responsible for inheritance, which he called gemmules, were collected from the body's cells and transmitted via sperm and egg. Weismann was a staunch advocate of Darwinism, but he was convinced that there was a fundamental distinction between the somatic cells that made up the body's tissues and the special cells called germ cells that gave rise to gametes. Any changes to the "plasm" of somatic cells could therefore play no part in heredity. To demonstrate that changes to the body of an organism are not inherited, Weismann cut off the tails of hundreds of mice and followed their offspring for five generations, each time removing the tails. Not once were any offspring born without tails.* Any notion that "acquired characteristics" could be inherited, as in the pre-Darwinian theory of evolution proposed in the early nineteenth century by Jean-Baptiste Lamarck, could no longer be sustained.

In Weismann's view, then, somatic cells are irrelevant to evolution. They are destined to die with the organism. But germ cells beget more germ cells – there is an unbroken line of germ cells (the *germ line*) down through the generations. It's often said that the germ cells are thus immortal, although that's an odd formulation – by that definition, we are all immortal simply by virtue of being able (if indeed we are) to produce offspring.

* * *

In the story of how to make a human "the natural way", the fertilized egg is often portrayed as the end – at least, until the happy day that the baby emerges. All our traditional stories of people-making rely on that quantum leap from fertilization to birth. The dire moral warnings about pregnancy that loomed over adolescence (and in some cultures still do) make this the equation: bring together sperm and egg and *you'll get a baby!* It's a warning (sometimes needed, for sure) to experimenting teenagers, but becomes more like a promise in the narrative of IVF: to make that longed-for baby, all you need to do is unite the gametes. And if it doesn't turn out that way, *something has gone wrong.* There is a single and inevitable road from fertilized egg to infant, and anything else is an aberration.

This is misleading. To put it starkly, most acts of non-protected penetrative sexual intercourse do not produce a baby – and when I say most, I mean 99.9 per cent. Even most fertilized eggs do not become babies – about 2 to 3 in 10 confirmed pregnancies abort

---

* It's not obvious that this rather harsh experiment was needed at all, for as Weismann himself pointed out, "Jewish boys are not born without foreskins".

spontaneously in miscarriage, but even those figures mask the 75 per cent or so of fertilized eggs that never get to the point of registering as a pregnancy at all, either because they don't develop into a multi-celled embryo or because the embryo fails to implant in the uterus. That's a puzzling thing about humans: we are unusually poor, within the animal kingdom, at reproducing. You have to wonder whether all the attention we give to sex is because we are so spectacularly bad at getting results from it.

Even to say "bad" is perhaps to collude with the moral imperative of the fertilized-egg-to-infant story; let's just say that we are an anomaly, for reasons imperfectly understood. This calls into question the idea that sex really is "for" reproduction, as some religious moralists insist. If we were inclined to see procreation as a divine gift and imperative, one would at least need to grant that God expects us to have a heck of a lot of rehearsal.

The baby grows, of course, from a fetus: even children's books tell us that. But in the common view the fetus *is* simply a baby – a person – that has not yet fully developed. Its proportions might be a little odd, its limbs blunter, but it is recognizably human. The classic images made in the mid-1960s by Swedish photographer Lennart Nilsson and presented in the book *A Child Is Born* (1965), have defined the view of our *in utero* existence ever since. They show the fetus floating freely in space, often lacking even an umbilical cord, like the iconic image from Stanley Kubrick's *2001: A Space Odyssey* three years later. Perhaps this "child" even sucks its thumb. But these images were actually made by artful arrangement of aborted fetuses – they were not in fact living organisms at all, much less *in utero*. They were curated to tell a reassuring story. (At least, so it might seem until you realize that it's a story in which the mother has been edited out.)

By the time a fetus looks even vaguely human (which is what, loosely speaking, distinguishes it from an embryo), most of the important stuff has happened. Most of the dangerous hurdles have been cleared. And most importantly, the developing organism is already anthropomorphic, relieving us from any need to grapple with the strangeness of an entity evidently made of cells, which we might want to call human but would struggle to justify that intuition.

Yet it is the early embryo that reveals the true versatility, the genius, of our cells – and the unfamiliarity of the moment when those cells are not merely what we are made of, but what we are.

It might surprise you to discover – it surprised me – that when a woman first has a fertilized egg (a zygote) inside her body she is not technically pregnant. This is not some perverse biomedical fine print; it simply makes no sense to see things otherwise. A pregnancy test would show nothing, nor will it for the first four days or so after fertilization. The zygote divides by mitosis into two, then four, then eight cells and so on, and at this point these cells can form all the tissues needed in the embryo: they are called *stem cells* and are said to be *totipotent*.

In other words, every one of these cells could potentially become a separate embryo. In the early days of embryology, that was by no means clear. The German zoologist Wilhelm Roux thought, for example, that cells are headed towards different fates from the first division of the zygote. In 1888, he reported experiments on frog embryos at the two and four-cell stage, in which he destroyed one of the cells by lancing it with a hot needle. A single remaining cell from a two-cell embryo would then, he said, grow into a half-embryo, suggesting that it had even at that stage become assigned as the progenitor of that part of the body plan alone.

But Roux's method was flawed, because he could not detach the remains of the ruptured cell from the intact one. This debris interfered with the subsequent growth of the embryo. In the 1920s and '30s, German embryologist Hans Spemann performed a cleaner act of surgery on salamander embryos. By using a noose made from a single hair taken from a baby, he pinched early embryos in two and found that each of the resulting parts is able to grow into a complete embryo.* In effect, Spemann made identical twins by artificial means. Because he produced two genetically identical embryos from a single initial one, you could also call this a process of cloning.† Spemann and his co-workers used amphibian cells, because they are so large that the delicate manipulation could be done by hand – albeit an impressively steady one.

The ball of totipotent stem cells that is the human embryo floats freely in the fallopian tube (also called the oviduct), borne slowly towards the uterus. By day five, the embryo has become a ball of around 70 to 100 cells and has rearranged itself into a structure known as the blastocyst. By the time it arrives at the uterus, it has shed the protein coat called the zona pellucida that formed the protective shell of the original egg – it has "hatched" and is ready to implant.

The human embryo at around five days, called a blastocyst.

---

* In fact, as early as 1869 Virchow's student Ernst Haeckel showed that fragments of the embryos of jellyfish-like organisms called siphonophores can produce complete larvae. Siphonophores are, however, very unusual creatures, being not true jellyfish but colonies of small multi-celled organisms called zooids.

† This won't be the outcome for all vertebrates, however. Separate the cells of a two-cell mouse embryo, and in general only one of them will grow into a mouse – showing that in this case there are differences between the cells even at this extremely early stage.

That ball of cells is not exactly the nucleus of a person. Most of the cells of the blastocyst became the mere housing and life support. Some of them form an outer layer enclosing a fluid-filled void: these are trophoblast cells, comprising the tissue called trophecto-derm which will become the placenta. Others congregate into a clump on the inside, called the inner cell mass, which separates into the epiblast from which the fetus will grow, and the hypoblast that will eventually become the yolk sac. The epiplast consists of embryonic stem cells, capable of forming all the tissues of the body (but not the placenta): a capacity called *pluripotency*. Identical twins grow from two separate inner cell masses in a single blastocyst, whereas non-identical twins grow from two separate blastocysts, formed from distinct eggs fertilized by different sperms. Within a few days of implanting, the epiblast is covered in a layer of special-ized cells called the primitive endoderm, derived from the hypoblast.

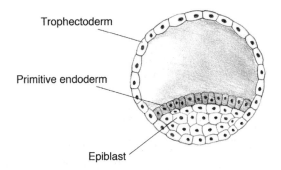

The human embryo at around day 10–11.

The fate of the embryo wholly depends on a successful implantation in the lining of the uterus. If this does not happen – which is the case around 50 per cent of the time – the embryo will be expelled in the menstrual cycle. Failure to implant is one of the common reasons why an IVF cycle does not work. No wonder, then, the division of labour in the blastocyst makes it seem that its priority is to those cells surrounding the epiblast, which won't be a part of the fetus at all. For without implantation, it's game over.

Implantation is a delicate and complex process involving a dialogue of hormones and proteins between the embryo and the cells of the uterine lining. In some ways it is more delicate and complex than fertilization itself. The placenta, for example, is made not just from the trophoblast layer of the blastocyst but also from tissues from the mother, called the decidua. The two types of cell, with different genetic makeup, have to work together to create a single, vital organ. Emotive and anthropomorphic metaphors suggest themselves, presenting implantation as an intimate collaboration between the tissues of mother and her "child". But one might equally choose to speak of the blastocyst "invading" the uterine tissue: one "organism" colonizing another for its survival.* Both are stories; neither is a neutral description of events (which story ever is?).

* * *

The best is about to come. Calling the part of the embryo fated to become the baby an "inner cell mass" is no euphemism: it really does seem to be a shapeless conglomerate. If we want to insist that baby-making is a miracle, what seems truly miraculous is not just that the inner cell mass makes a body but that, most often, it makes exactly the same type of body, with five fingers on each hand, with all facial features in the right place and fully functional, and with its battery of correctly positioned organs. It's no surprise that development of the embryo occasionally goes awry; it is astonishing that it does so rather rarely.

---

* If you think that seems a harsh way to put it, bear in mind that some of the important biochemical processes that take place in the placenta involve genes that seem to have originated in viruses.

When embryos start off as single cells, they have no plan to consult. Cells are programmed to grow and divide, but it isn't meaningful to think of a human being as somehow fully inherent in a fertilized egg, any more than one can regard the complex convolutions of a towering termite mound as being programmed into each termite. The growth of an organism is a successive elaboration of interactions within and between cells: a kind of collaborative computation whose logic is obscure and convoluted, and the outcome of which is incompletely specified and subject to chance disturbances and digressions.

In this way, the job evolution has devised for those formative cells is an architectural one: a challenge of coordination in time and space. They have to move into position, to acquire the right fate at the right time, and to know when it is time to stop growing or to die.

Developmental biologists talk of this as "self-organization". It could make the process sound quasi-magical, calling as it does upon the image of the cell as an autonomous being with aims and purposes. But many of the rules are now broadly understood.

Two key factors are at work. First, as the cells divide and multiply, they take on increasingly specialized roles, a process called differentiation. Thus, totipotent cells in a two or four-cell embryo become trophoblasts or the pluripotent stem cells of the epiblast. The latter go through further stages of differentiation that ultimately produce the specialized cell types found in muscle, skin, blood and so forth. We will see shortly how that happens.

Second, particular spatial arrangements may arise from cells actively moving through or across the growing organism or organ, or becoming sorted into clumps of different cell types by preferential stickiness, often between cells that are alike.

That cells have adhesive qualities joining them into aggregates was suggested in the 1890s by Wilhelm Roux. He was also able to disrupt frog embryos by vigorous shaking, which separated them into single cells. He found that those cells would join back together, which he attributed to some kind of attractive force.

Such "disaggregation" experiments were taken further in the early 1900s by marine biologist Henry V. Wilson, who found that sponges kept for a long time in an aquarium became "loose" and could be teased apart into individual cells. He achieved the same thing in fresh sponges by the simple measure of squeezing them through a piece of silk, which acted as a sieve that separated the cells. Again, those cells would reassemble if brought into contact to regenerate a living sponge. It was like a recapitulation of the evolution of primitive multi-celled organisms from colonies of single-celled ones (see the First Interlude, page 86). When Wilson did the experiment with different species of sponge, he found that cells from the same species would stick together selectively. Ernest Everett Just discerned in the 1930s that the reason for this selectivity had something to do with the cell membranes. The truth is that cells adhere via protein molecules protruding at their membrane surface (especially those belonging to the class called cadherins), which will bind to one another discerningly.

This notion of "tissue affinity" was developed around the same time by the German-American embryologist Johannes Holtfreter. In 1955, he and Philip Townes studied how the cells of amphibian tissues that had been disaggregated by exposing them to alkalis could reassemble from solution. Holtfreter largely outlined the concept of cell sorting that allows tissues of several cell types to adopt particular structures and arrangements.

The process of body formation (morphogenesis) is orchestrated by genes, and no wonder then that genes have been attributed such determinative power. Some researchers have made more apt comparisons to a musical score: genes tightly constrain but do not fully prescribe the performance. This is still a limited metaphor, because you can look at the score and figure out (if you're a musician) pretty much how things will go. Not so with genes. Sometimes it is better simply to tell the story as it is, as simply as you can, rather than trying to pretend it is some other story.

Morphogenesis literally means shape-formation, but equally it is a question of cell specialization: the embryonic stem cells gradually lose their versatility as they divide, becoming geared instead to do the task of specific tissue types. Heart muscle cells must execute synchronized beating, pancreatic cells must secrete insulin, the nerve cells of the eye's retina must respond to light, and so on. This happens not by cells gaining new properties, but rather by narrowing the possibilities inherently available to them by shutting down genes that aren't needed. That's what differentiation is all about.

The cells must know how and where to switch genes on and off as differentiation proceeds. How do they know? The cues come from the other cells and tissues around them.

Some of these signals are delivered as chemical messages, which, diffusing through the mass of cells, serve to define a kind of spatial grid that lets cells know where they are in the overall embryo and thus what their fate should be.

Imagine that a cell, or group of cells, at one place in the embryonic mass switches on a gene that produces some protein. And suppose that this protein can diffuse out of the cell, like water leaking out of a paper bag, and into other cells. Then the concentration of the protein throughout the embryo varies gradually from place to place, being greater nearest the cells that produce it and slowly diminishing with distance. If you could measure the protein concentration, you'd have some notion of where you are in the embryo relative to the source cells. You'd be able to sense your *position*. Think of it in the same way as finding your way to the kitchen of a large house by following the smell: the stronger it is, the closer you are.

These "position-marker" proteins are called morphogens, and cells are able to "sense" their concentrations. Morphogen concentration gradients allow regions of the embryo to become distinct from one another.

To see how this can work, let's forget the human body for a moment and look at the development of a simpler embryo: that of the fruit fly. This humble creature became the paradigmatic representative of "complex life" in the early twentieth century, when its robustness and ease of breeding made it the ideal subject to study the mechanisms of genetic inheritance – an art of which Thomas Hunt Morgan was the master. There are, of course, substantial differences between humans and fruit flies, extending to their genetic and developmental fine print. In particular, fruit-fly embryos, unlike those of mammals, are not initially clusters of separate cells at all. Once fertilized, the ovoid fly egg starts to replicate chromosome-carrying cell nuclei, but just accumulates these around the edges of the egg. The nuclei only acquire their own cell membrane once the embryo has amassed 6,000 or so. This lack of cell membranes in the early embryo makes it particularly easy for morphogens to diffuse through it.

One simple way that gradients of diffusing molecular morphogens can mark boundaries is to think in terms of concentration contours. A contour denotes a *threshold*: a point where the concentration exceeds a certain value.

The fruit-fly embryo acquires its initial pattern features from morphogen threshold concentrations. Pretty much the first thing it does is to define which end will become the head and thorax, and which end the abdomen. In other words, the embryo acquires a front–rear axis. That is defined by a morphogen protein called bicoid. At the tip of the "head" (so-called anterior) end, the embryo produces bicoid, and this begins to diffuse down to the rear (posterior) end. The concentration falls smoothly from the anterior to the posterior end. Where it exceeds certain values, the bicoid protein will bind to the DNA within the embryo and activate other genes with vivid names like hunchback, sloppypaired 1 and giant (typically named because of the developmental defects that mutations in the genes can produce). How this switching occurs is complicated, not least because it also seems to depend on a gradient of another protein called caudal that diffuses from the opposite (posterior) end. But the outcome is that the embryo becomes quite sharply segmented into regions where different genes are expressed or not. Thus the uniformity of the embryo is destroyed: an anterior–posterior axis is established, along with the segments that will develop into the fly's head, thorax and abdomen. It seems that similar gradients cause segmentation of the neural tube of verte-brates: the tissues that will become our brain and spinal column.

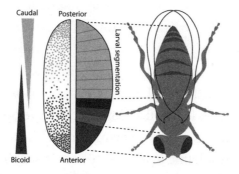

Gradients in the concentration of proteins bicoid and caudal from opposite ends of the fruit-fly embryo switch on genes at different positions that cause segmentation of the body plan.

Other diffusing morphogens produce other kinds of gradient, defining different axes of the emerging body. For example, a protein called dorsal is involved in setting up the top-to-bottom (dorsoventral) axis of the fruit-fly embryo that distinguishes the region that will become the back (where the larva will ultimately grow wings) from that which will become the belly. In each case, the gradient thresholds may turn particular genes on and off in a series of elaborations that begins with the crudest determinants of shape – the front/back and top/bottom axes, say – and works its way to the fine details.

The idea that chemical concentration gradients might control the development of embryos was first proposed at the start of the twentieth century by Theodor Boveri. By producing a chemical patterning signal that spreads into the rest of the embryo, one cell can determine the fate of other cells nearby. In 1924 Hans Spemann, together with Hilde Mangold, called such groups of cells "organizers".* Mangold transplanted groups of cells in amphibian embryos from one position to another and saw that they could induce the development of "out of place" features.

The British biologists Julian Huxley and Gavin de Beer verified the idea of organizers in the 1930s by manipulating the embryos of birds. They proposed that Spemann's organizers create "developmental fields" of some kind that influence the course of development. Spemann had imagined this "field" as something like the magnetic or electrical fields of physics, but Huxley, de Beer and their contemporaries in this fledgling field of developmental biology suspected that the agent was a chemical one. The notion that these organizing centres define a sense of position within the emerging body plan through the action of morphogen concentration gradients was crystallized in the late 1960s by biologist Lewis Wolpert.

There's a crucial part of this story that I've skipped over so far. The patterning of the fruit-fly embryo is kicked off by the

---

* Mangold was Spemann's graduate student at the University of Freiburg and did the experimental work behind this discovery. She died in a domestic gas explosion while heating milk for her baby before the paper she wrote with Spemann on organizers in embryogenesis was published, and therefore could not share the Nobel prize that Spemann won in 1935 for the work.

production of the bicoid protein at the anterior tip of the egg. But what causes that production in the first place? How does the bicoid gene know it is at the anterior end?

The answer is that "mother tells it". While the unfertilized egg is attached to the follicle of the mother fly, specialized cells called nurse cells deposit material needed to make bicoid – specifically, the RNA molecules that mediate the gene-to-protein conversion – into the anterior tip of the egg, so that developmental patterning is all ready to go when the egg is fertilized. Right from the outset, cells in the embryo are dependent on other cells around it to know what to do. It's for similar reasons that a fertilized human egg can't develop fully in isolation, if cultured in a test-tube. Implantation in the uterus wall is needed to give it a "sense of up and down". Ectopic pregancies (within the fallopian tubes) show that such a signal doesn't *have* to come from the uterus, however, and we'll see later whether there might be other ways to do produce it *in vitro*.

This is why it is strictly incorrect to say – although it often *is* said – that all the information needed to grow a human being is in the genome of the fertilized egg, which is in turn supplied by the gametes that combined to make it. You could say that the human embryo also needs *positional information* supplied by its environment – specifically by the uterus lining. Furthermore, any particular cell in the developing embryo depends on receiving information from the surrounding cells in order to keep embryo growth on track. As the transplantation experiments of Huxley and de Beer showed, if you mess with that information then you screw up development, despite the fact that every cell retains its complete "genetic programme".

Embryo development is thus not encoded from the outset in the genome, as if in some blueprint or instruction book. It relies on a precise expression of genetic information in time and space, which in turn depends on the proper coordination of many cells (including maternal ones) and is subject to chance events during the execution. To understand embryology and the growth of complex tissues and organisms, we shouldn't imagine that we will find a set of instructions packed like a homunculus inside the zygote. Rather, we will need to discern and interpret the patterns of *information flow* (and the various sources of that information) as the process unfolds.

It's rather as if the genome is a list of the words that feature in a book, but you need *other* information to put the words in the right order so that they become more than just assemblages of letters and may take on meanings. Those meanings are not inherent in the words themselves but may be determined by the words nearby, by allusions and interactions that leap from one part of the text to another – by context. Again, there is no perfect metaphor for illustrating how genes work in building an organism; doubtless this one would collapse too under pressure, so use it gently.

* * *

I won't explain in detail how human embryogenesis differs from that of the fruit fly, but it's worth understanding one of the most basic distinctions. For the human body doesn't simply emerge imprinted on the inner cell mass of the blastocyst like stripes on an embryonic zebra.* Rather, the cells in the embryo move around,

---

* These too are thought to be a pattern made by morphogens – in this case, by the interaction of at least two morphogens, which control the production of pigment in skin cells. A theory of stripe formation among diffusing morphogens was proposed in 1952 by the British mathematician Alan Turing, and his scheme is now thought to apply to many patterning processes in animals and plants, including the quasi-regular spacing of hair follicles and the formation of ridges on the canine palate. It too invokes diffusion as a means of mapping out positional information, but this time in combination with chemical reactions involving the two morphogens, which introduce feedback into the equation and give rise to more elaborate patterns. Turing's scheme showed how the basic ingredients of a patterning system can, by their intrinsic nature (their rates of

and the tissues grow, buckle and fold, to shape the body. It's a highly active process, a kind of auto-origami happening in parallel with the appearance of distinctions between cell types. The first stage of this process, which for humans occurs around day 14 after fertilization (around the time that a pregnant woman might first notice a missing menstrual period), is called gastrulation. Some scientists regard *this* as the point where a mass of cells begins to produce an organism: as the beginning of personhood.

There is a lame joke that scientists still seem to find amusing about how, if a physicist were to study the cow, she would first simplify the question by approximating it as a sphere. It is rendered all the lamer by the fact that this is not so far from how nature approximates the human body – or the bovine one for that matter – in the first instance. For the most rudimentary idealization of our body might run thus: an inner tube for digestion from mouth to gut to anus, an outer layer of skin to create a boundary, and everything else packed into the space in between. At one end we'll put the head – the anterior – and at the other end is of course the posterior. Gastrulation creates a structure very much like this (the word actually means "gut formation"). In some creatures, such as species of worms and molluscs, it really is that simple: gastrulation folds the embryo into a sort of fat tube or doughnut shape in which an inner tube connects mouth to anus, and the job is nearly done at a stroke.

For humans, it is rather more complicated. The embryo develops a central groove called the primitive streak, which will become the axis of the backbone and central nervous system: the beginning of the aforementioned neural tube. The subsequent folding is not easy to describe in words, but it creates the crescent-shaped structure

---

reaction and diffusion), give rise to a pattern with a characteristic length scale and type of symmetry.

that will become the fetus, connected to a yolk sac (involved in early embryonic blood supply) and attached to the placenta via the umbilical cord. The key point is that initially this gastrulated human embryo develops distinct types of tissue: its cells lose their pluripotency and start to specialize. The innermost layer, which will form the lining of the gut, is called the endoderm ("inner skin"). The outer layer, or ectoderm, generates the surface layer of the skin as well as the brain and nervous system. Between these layers is the mesoderm ("middle skin"), which is the primal fabric of the inner organs and tissues: the heart, kidneys, bone, muscles, ligaments and also the blood. At this stage, some of the embryonic stem cells are also set aside to become the germ cells: the precursors to the gametes (eggs and sperm).

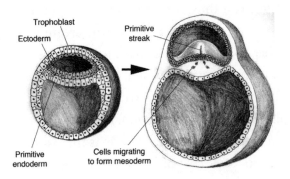

Gastrulation of the human embryo and
formation of the primitive streak.

And there you have it: the schematic human body, its cells launched on their road to specialization. The rest is refinement. For example, some neural cells in the head region develop (around week five of gestation) not into neurons but into the retinal cells of the eye. Some cells don't differentiate where they first sit, but actively migrate through the embryo to where they need to be – we saw that the primordial germ cells do this. The sex organs develop identically at first in both sexes, becoming female organs unless triggered into the structures of the male if the cells have a Y chromosome instead of a second X. On the Y chromosome sits a gene denoted SRY, which controls other genes needed to develop male characteristics.

All of this refinement happens through cell dialogue. Molecular messages pass from cell to cell, each at the proper stage of development, so that cells get assigned their roles in collaboration with their neighbours. "The parts of each organ help the other parts to form," explains cell biologist Scott Gilbert. It is because organoids like my mini-brain lack this information from surrounding tissues that their development – their morphology – is imperfect. To make a body or even a mature organ, cells need community.

*   *   *

The idea that genes involved in development interact and control one another via diffusing morphogens is only a part of the story of how embryos take on their form. The distinctions between cell types initiated by such signals become permanently imprinted on the cells as they develop into different tissues.

How can that be, if they still all share the same genome?

That problem was recognized by Thomas Hunt Morgan and others in the early days of molecular genetics, but no one really knew how to address it. So it was largely put to one side. The discovery of DNA's genetic code in the 1950s and '60s all but eclipsed the question, seeming as it did to promise an underlying simplicity in the way cells function. Already in 1941, Morgan's former student George Beadle, along with biochemist Edward Tatum, had shown that genes (whatever *they* were – it wasn't yet clear) encode protein enzymes. This became understood to mean that each gene has a unique corresponding protein. The key question was then *how* a gene made a protein. Crick and Watson's double helix, zipped together with information-bearing nucleotide bases, seemed to deliver the answer: DNA carries the coded plan, and RNA and ribosomes are the machinery that does the translation.

But how do you get from a protein to the phenotypic effect of a gene on the unfolding organism? That wasn't at all obvious. By the 1960s, the general idea was that genes *act* in some vague way to dictate the developmental programme, which was then envisaged merely as "an unfolding of pre-existing instructions encoded in the nucleotide sequences of DNA", as American biologist-cum-historian-cum-philosopher Evelyn Fox Keller has put it. According to the French biochemist Jacques Monod, as far as gene action is concerned, "what's true for [the bacterium] *E. coli* is true for the elephant." What seemed to matter was establishing the common basis by which gene becomes protein. Somehow the rest – meaning the living organism itself – followed. Which would be all very well, if *E. coli* looked like an elephant.

In this picture, then, the answer to the question of development must reside in the gene sequence, and the ultimate goal of biology becomes the decoding of that sequence. This picture has been burnished for a remarkably long time, culminating in the Human Genome Project, which began in the 1990s and announced the almost complete sequencing of the human genome between 2001 and 2003.* The objective was simply to get the code, which took

---

* This "human genome sequence" creates a lot of misunderstanding. It prompts the question "whose sequence?", because the whole point is that our individual differences are largely genetically determined. But our genome sequences are about 99.9 per cent identical – it is just the small differences in that remaining 0.1 per cent or so of genetic code that make us unique. What's more, many of those differences involve gene mutations widely shared among different people, albeit in different permutations. And these different variants of a specific gene encode the same function – supplying the blueprint (here, yes!) for a protein enzyme, say. It's just that the corresponding proteins may differ slightly in their molecular structure and activity. So a universal entity called "the human genome" is a meaningful notion, even if it is a little fuzzy.

on the status of the "fundamental" information directing all biological activity. Meanwhile, genetics more broadly looked for correlations between genes and phenotypic outcomes. Exactly how and why genes exert their effects was a question long bundled up in the vague concept of "gene action" that, as Keller says, allowed scientists "to get on with their work despite almost complete ignorance of what that 'action' consisted of." There was an implied hierarchy of causation in which genes were paramount, as reflected in Nobel laureate David Baltimore's comment that the development of an organism involved the "greasy machines" of the cell directed by the "executive suite" of the genome. (Engineers are very familiar with this kind of prejudice.)

The resulting view was that development was a kind of painting by numbers of the plan in the genome. For a complex organism like us, that left an awful lot of instructions to be packed into the genes. As the Human Genome Project got underway, biologists estimated the number of genes the project would find as being somewhere between 140,000 and a lower limit, proposed by a few bold souls, of 26,000. Most put the figure at around fifty to seventy thousand.

The answer turned out to be 23,000.

This is often presented as a sobering example of how experts can get things wrong. It's certainly that, but rarely does anyone identify the real moral: that the genome doesn't work the way it was thought to.

Zoologist Fred Nijhout is one of the few to have come properly to terms with the implications. "A more balanced and useful view of the role of genes in development," he says, "is that they act as suppliers of the material needs of development and . . . as context-dependent catalysts of cellular changes . . . they simply provide efficient ways of ensuring that the required materials are supplied at the right time and place." They are less like Baltimore's executive directors, and more like stewards guiding a crowd. It's no coincidence that Nijhout sees things this way, because he is an expert on the genetics of butterfly wing patterns, where it is clear that just a few genes, creating interacting fields of influence through the diffusion and spreading of morphogens, can generate a startlingly diverse array of patterns and forms, dictated by the details of how the genes are expressed in time and space. It's somewhat meaningless, in such a situation, to say what a gene does (beyond "encode a class of proteins") without specifying where and when it acts.

The view now emerging is that a relatively small number of genes is able to generate the complexity of the human form, with its many different tissue types so precisely arranged and coordinated, because they act in networks that produce distinct patterns of gene expression varying over time. With 23,000 genes, the number of possible *networks* of influence is astronomical.

How do genes acquire and change their patterns of behaviour? The control, activation and silencing of genes in different cell types and at different stages of development is called epigenetics. The word literally implies something *additional to* genetics, but what it really connotes is that the observable outcome of genetic activity – the phenotype, such as the tissue type of a cell – isn't determined by the genotype (that is, which genes are present), but by the question of which genes are active. Epigenetics is all about how genes become modified to alter whether, or how much, they are expressed.

There are several ways in which this can happen. One is by the attachment of molecular "tags" to the respective genes, which might act as markers that deter the machinery of transcription, suppressing gene expression. Some genes can be switched off, for example, by proteins that stick a so-called methyl group – a carbon atom with three hydrogen atoms attached – onto DNA bases in the gene, which forms a sort of "shield" that protects the gene against being transcribed and translated into a protein. Another molecular mechanism of epigenetic regulation involves chemical changes to the histone proteins around which a stretch of DNA is wrapped in the chromosomes.

Harder to understand than this attachment of "leave me alone" labels to genes, but equally important for epigenetics, are processes that involve the packaging of DNA in chromosomes. Remember that the combination of DNA and histone proteins in chromosomes goes by the name of chromatin. This stuff is rather systematically coiled up and stowed when the chromosomes are in the compact form (typically X-shaped) found in dividing cells. At other points in the cell cycle, chromatin can be unwoven and loosely strewn, in which case the transcription machinery can get to it more easily. So how "active" genes are can depend on how the corresponding regions of the chromosomes are packaged.

An example of this epigenetic gene regulation happens in female cells, which contain two copies of the X chromosome, one passed on from each parent. If both of them were active, they would produce more proteins from this chromosome than the cell needs, and that would cause problems. So one X chromosome is inactivated early in the development of the embryo. The choice of whether to silence the maternal or paternal X chromosome is made by each cell at random and then passed on to daughter cells when they divide. The result is that females end up with an equal blend of two types of cell throughout their body. This process of X-chromosome inactivation was first identified by geneticist Mary Lyon in the 1960s. It took many years to figure out how "X-silencing" occurs, but we now know that it involves a gene that switches on a series of events resulting in the packaging of the inactive chromosome into a tight bundle, inaccessible to transcription. All the genes are still there and are faithfully copied and passed on when cells divide, but the shape of the chromosome keeps them hidden.

Some epigenetic changes to DNA that regulate gene activity happen automatically as cells divide and mature: each cell type will have its own characteristic pattern of epigenetic modifications. This too is why development of a fertilized egg into an embryo and then a mature organism isn't exactly just a matter of reading out a genetic programme. It involves a continual, ever-changing process of epigenetic editing of that programme, taking place through time and space.

*   *   *

In the mid-twentieth century, British embryologist Conrad Hal Waddington offered a metaphor for the process of epigenetic cell differentiation. He imagined cells in the early embryo traversing a landscape of possibilities: they begin their journey at a mountain peak and descend into valleys that branch like the channels of a river. At each branching point, the cell (more properly, the lineage of dividing cells) makes a decision about its subsequent fate: to become a progenitor of lung or heart, say. The consensus was that, once a lineage has descended into a valley, it can never reverse direction and go back uphill again.

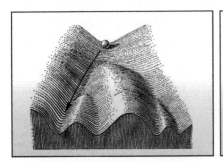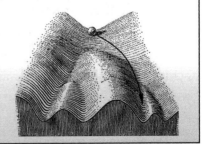

The Waddington landscape. The balls represent the trajectories of different cell lineages, which begin in the single valley of totipotency as the zygote first begins to divide, and end in valleys representing different types of mature, differentiated cell.

Differentiation begins early. Indeed, it has happened even in the pluripotent embryonic stem cells of the epiblast, which have lost the totipotency of the earliest cells made from the dividing zygote. Already some of those first cells have been directed down the valley in Waddington's landscape that leads to a placenta or a yolk sac, not to a part of the fetal body. The cells that make up the three layers of endoderm, mesoderm and ectoderm in the gastrulated embryo have undergone a further degree of differentiation, a further narrowing of choice.

It's because of this specification of cell "fates" early in embryo-genesis that the germ cells need to be formed so soon. Evidently a barely formed embryo doesn't yet "need" eggs or sperm – but it must put aside the cells from which they will grow before they lose too much of their pluripotency. The germ cells, after all, have to make a totipotent zygote, so it won't do if their chromosomes have already been heavily modified and silenced. Germ cells *do* have some epigenetic silencing of genes, although this too must be stripped away when the gametes combine to make a totipotent zygote.

This special dispensation for germ cells aside, epigenetic changes appear to be one-way. They partly account for how our body tissues remember and maintain their identity as they grow: why skin cells divide to produce more skin cells, and don't spontaneously become muscle cells or stem cells. In other words, cell replication is some-what more complex than merely a matter of copying the chromosomes. It's necessary also to copy the epigenetic chromo-somal modifications that give the cell its identity.

What this means is that each cell in our bodies, like each one of us, has a lineage: an ancestral history that starts with the zygote – and, except for a handful of germ cells (if we have children), ends in the grave. A liver cell has arisen from an embryonic stem cell via a succession of ancestors with intermediate characteristics, reflecting an ever greater specialization and loss of versatility. This notion of cell lineages was first articulated clearly by August Weismann when he drew up his fundamental distinction between somatic ("mortal") and germ ("immortal") cells.

When we tell the story this way, a new possibility becomes apparent. In cells during development, just as in organisms during evolution, genes can change. Every time a cell divides, there is a chance that some of the three billion base pairs in the genome will be miscopied – that the daughter cells will acquire mutations. Cells put a great deal of effort into avoiding such mistakes, employing a kind of molecular proofreading to check for errors in replication. All the same, the numbers are so vast that mutations are inevitable. It's estimated that distributed within the chromosomes of our 37 trillion or so cells are about ten thousand trillion genomic mutations.* Every one of our genes experiences somatic mutations at some point in our lives.

Most of these mutations, fortunately, don't matter – they don't affect a gene's ability to do its job(s). But some have consequences. Most notoriously, gene mutations can lead to the cell dysfunctions that cause cancer (see page 130). Even potentially harmful mutations, though, might not matter if they happen late in development and so appear only in a few cells. Somatic mutations that arise during early embryo growth, on the other hand, may be passed on to all subsequent cells in that lineage, making the developing body a patchwork or "mosaic" with slightly but perhaps

---

* It's likely that the incidence of mutations – the error rate – is in fact "optimized" during evolution, since mutations are precisely what evolution "needs" to work at all. In other words, the error rate isn't necessarily kept as small as it possibly can be but is finely tuned via the efficiency of the proofreading mechanisms to ensure that the resulting genetic variation is as beneficial as it can be. In any case, these variations from cell to cell in our bodies are why, strictly speaking, the idea that we each have a single, unique genome sequence is wrong.

significantly different genomes. There are many diseases related to "mosaicism", of which cancer is just one class. Somatic mutations leading to mosaicism are particularly common in brain neurons, and they are thought to be responsible for a range of brain and cognitive disorders, including some types of autism. Even benign mutations can manifest themselves in outward appearance (that is, phenotypically): for example, causing striped skin pigmentation called Blaschko's lines or the red skin blotches called port-wine stains.

One particularly unusual but very rare kind of mosaicism happens when a cell in a male embryo fails to pass on its Y chromosome to the daughter cells, which, inheriting only the X, then develop as "female cells" by default. This can lead to a mixture of male and female characteristics in the embryo. Rare it may be, but this condition serves to remind us of the cell's autonomy. Even in a body "meant" (to judge from the zygote) to be male, there is no global command that cells obey, and the "feminized" cells will feel no obligation to conform to the nature of their "male" neighbours.

Genetic variations along a cell lineage are, therefore, random. Epigenetic modifications that give rise to different cell types and tissues, on the other hand, are generally systematic and preordained in the genes – not in the sense that they will happen come what may, but that they are destined to be a part of the developmental programme so long as it proceeds without mishap. Some epigenetic changes aren't preordained at all, though. They may take place in response to the contingent environment of a cell or organism, including unpredictable events arising from randomness within the network of interacting genes themselves. This is one reason why identical twins, who have the same genomes, may look rather different later in life. They have different environmental nudges, and this affects the epigenetic programming of their genes. Some dietary chemicals such as curcumin (found in curry spices) and resveratrol (in grapes) seem to have epigenetic effects on the folding-up of chromatin in cancer cells, while deficiencies of folate (a chemical in pulses and grains) can alter epigenetic patterns of methyl attachment to DNA. (Whether this means that red wine and tikka masala protect against cancer is another matter.) Drugs and pollutants can also act, for better or worse, via their influences on the epigenetic (as well as genetic) programming of cells.

Given that Waddington proposed the idea of an "epigenotype" – which he called a "whole complex of developmental processes between the genotype and phenotype" – in 1942, it is a little odd that epigenetics has been portrayed in recent years as a field that is "revolutionizing" biology. Perhaps that's just how it looks if you're starting from a simplistic view in which cells are nothing more than player-pianos orchestrated by their punched-hole genetic scripts – that's to say, if you had a faulty story to begin with.

All the same, it is only in the past several decades that we have had a more detailed understanding of how epigenetics works at the scale of cells and molecules. There are still many holes in that understanding. Some researchers now talk in terms of an epigenetic code that imposes itself on and modulates the "all-powerful" genetic code. But epigenetics is a dynamic process, for which a "code" might be the wrong metaphor. Sure, there may be an epigenetic signature characteristic of, say, a fibroblast cell. But the epigenetic status of a human being is constantly changing and depends on our personal history.

It is the recognition of contingency in a cell's epigenetic state that underpins the *real* revolution in biology. For, as the growth of my mini-brain from skin cells attests, the specification of cell fate is *not* irreversible. If we cling too strongly to the evolutionary metaphor of cell lineage, that sounds crazy: it's like saying that you could be transformed back to a pre-human Australopithecus (more properly, to the common ancestor we *Homo sapiens* shared with that early hominid).

But for cells, such things *are* possible. The history of our flesh can be reversed and revised, and this completely transforms the possibilities for what it can become – and what we might do with it. We will see later how this can be achieved.

* * *

It should come as no surprise that there's plenty of contingency and circumstance involved in the way genes, epigenetics and cell interactions combine to create a human being. Of course the environment can play a major and perhaps even catastrophic part. Drugs (licit and otherwise), alcohol, hormones and environmental contaminants entering the mother's bloodstream during pregnancy can disrupt the process, for example, in ways that are transformative to the embryo or fetus.

We might tend to imagine that this is just a matter of a "plan for a person" that either proceeds as it should or gets thrown off course. But I will end this chapter by looking at one more way in which a simplistic picture of the person being a kind of "read-out" of the genes in a zygote can be profoundly misleading. The person – their body, their chromosomal inheritance of propensities – is not so easily condensed into a single type of cell. For just as human societies can be diverse, so can the cell societies that comprise a human individual.

Non-identical twins, for instance, may each have a mixture of red blood cells from both twins. Red blood cells are unique in the human body in having no chromosomes: they are produced not by cell division but by transformation of a special kind of cell in bone marrow. They fall into particular general classes – blood groups – depending on the chemical structure of protein molecules on the surface of the cells. Normally, each individual has red blood cells of a specific blood group, but twins can have a mixture of each twin's blood group.

This was first discovered in non-identical twinned cattle calves by the American biologist Ray Owen in the 1940s. In 1953, British physicians Ivor Dunsford and Robert Race found a similar case of two distinct blood groups in a human, a patient denoted Mrs McK who was tested before donating blood. Mrs McK had no living twin, but she told the puzzled doctors that she had had a twin brother who had died at three months old. The mixture of blood types here come from the fact that twins share a blood circulation system in the uterus, and so may exchange blood-forming cells that continue to produce blood long after birth and perhaps for a lifetime.

This presence of cells from more than one "biological individual" persisting and carrying out their biological function in a single organism is said to make it a *chimera*. Robert Race coined the term in describing the case of Mrs McK, admitting that he was simply looking to give his paper a catchy title.

People can be chimeric in much more dramatic ways than this. Their entire bodies can be patchworks of cells that seem to come from two different people. One way this can arise is by the fusing, at a very early stage of development, of non-identical twin embryos *in utero*. In a demonstration of how our cells can adjust to "unforeseen circumstances", these fusions can give rise to an anatomically normal individual whose cells got their genetic material from two different pairs of gametes: they are said to be tetragametic. This can happen even if the embryos that fused were of different sexes: the reproductive organs will then be decided by which set of cells in the merged embryo happens to produce them. But the chimeric person's body as a whole is not specifically gendered one way or the other in terms of the usual XX/XY chromosomal distinction: it is a bit of both.

Chimerism can also arise from exchange of cells between an embryo and the mother carrying it. These two "individuals" are conjoined via the placenta, which, as I explained earlier, is a mixture of cells from both of them. But the placenta is a rather leaky barrier. So cells from the mother can become incorporated into the embryo and fetus, while the developing child's cells may enter the body of the mother.

In fact, some degree of exchange, known as microchimerism, is normal. Many women who are pregnant with sons, for example, acquire some cells with Y chromosomes. What surprised researchers when this exchange came to light in the 1990s is that these fetal cells may persist and remain active, albeit at a low level, in the mother for many years after the child is born. But while microchimerism affects just a small proportion of the body's cells, a process like embryo fusion to create a tetragametic chimera makes a person who is genetically heterodox through and through: some organs and body parts come from one embryo, some from another.

If I were a mixture of the flesh of "two people", what would that make me? It is tempting to say that such an individual is indeed a mixture of two people in one. But that seems a profoundly odd and unhelpful way to look at the matter, for in what meaningful sense were those twin embryos "people" before they fused? These clusters of cells have only given rise to a single individual. This is just one of the ways in which quirks of developmental and reproductive biology undermine a simplistic determination to invest an embryo with unique personhood. A "person" is a higher-order concept, not to be reduced to genes or cells.

Still, our habits of thought and even our laws are challenged by these discoveries. DNA analysis of a tissue sample from a tetragametic woman may fail to confirm that her biological children are "hers", if the sample does not happen to share chromosomes with her gametes. Such cases have come to light through genetic testing to confirm maternity in applications for social welfare benefits in the United States, leading to harrowing accusations of false claims of parentage. Some of these cases have highlighted how strongly we invest notions of personhood and identity in the character of our flesh and genes. In his book *She Has Her Mother's Laugh*, science writer Carl Zimmer describes two such cases, saying that the discovery of their chimerism left these women with "haunting questions not only about their families but about themselves". One woman wondered if she was only partly the mother of her children, despite having given birth to them, and partly their aunt. "I felt that part of me hadn't passed on to them," said the other woman. As Zimmer explains:

> We use words like *sister* and *aunt* as if they describe rigid laws of biology. But despite our genetic essentialism, these laws are really only rules of thumb. Under the right conditions, they can be readily broken.

Yet I wonder. Not all cultures *do* use these words this way. It is common in Chinese society, for example, to call a close female friend of the family "aunt" even without any blood relation, and in the West "sisterhood" and "brotherhood" are widely used to express sympathetic bonds irrespective of sibling connections. Many cultures have a flexibility of familial relations that does not inevitably reduce them to blood and birth.

No, the problem here is not that biology destroys our traditional categories and concepts of human life, but that we too often now fall into the trap of imagining that biology can and should arbitrate on socially mediated questions of self and identity, family and kinship, sex and gender. Biology has a habit of declining that role, handing back (so I like to see it) the responsibility and saying, "you, not I, are the ones who care about these issues, so you must decide them for yourself."

THE HUMAN SUPERORGANISM

HOW CELLS BECAME COMMUNITIES

To insist that the embryo is "us" from its first instants is to some degree a displaced religious impulse. It announces a moment of creation, as profound and abrupt as the *fiat lux* of the Old Testament. Because, let's face it, the world *did* begin when you did, and it will end when you do: that's a universal, experiential human truth. Symmetry alone then seems to demand a beginning that is as abrupt and all-encompassing as the end – a moment, in those monotheistic traditions, when the soul enters the body to match the one when it leaves.

But this concept of the embryo denies the true wonder of our origin, and is another expression of the flight from flesh that has been going on for centuries. The assertion of the soul as an immaterial thing, pre-existing and eternal, is a pre-scientific attempt to deal with the incommensurability between the life we lead and the life in our cells.

For the latter is truly something to be astonished at. It is contiguous with the moment life first appeared on Earth. Life is passed like a baton between living things and is not created afresh with their own beginning. In arguments about abortion and embryo research, we talk about "when life begins", but that's not what we mean. *Life only began once*, around four billion years ago, and no one knows how. It continued in an unbroken thread from primal slime and algae to the oddly shaped metazoans of the Cambrian, through to the shrew-like ancestors of all mammals, and on to our apelike forebears walking upright and wielding stone tools, and finally – for this brief, glorious moment – here you are. Life is just passing through you, so enjoy it while you can.

The ambiguity, anxiety and angst that arise when we contemplate the life of the one-cell zygote and try to reconcile it with the human form in order to formulate laws and moral codes are consequences of our being assemblies of cells living in community. So it's worth considering how that came about.

* * *

If there was to be a competition for the least appealing organism in the world, slime moulds would be a strong contender. Bacteria get a bad press as mere "germs" to be expunged, but they also have a certain cachet too now that we know their presence in our gut is so beneficial and that they have such superpowers: metabolizing radioactive waste and oil spills, surviving in hot springs and so forth. Slime moulds, meanwhile, appear to be nothing more than their name suggests: a slightly disgusting smear of living matter whose purpose seems incidental to anything useful or inspiring in nature.

These organisms are members of a group called Mycetozoa. They are a type of amoebae, single-celled entities so "primitive" that for years microbiologists argued about whether they were closer to animals, fungi or plants. Modern genetic studies suggest that in evolutionary terms they are most closely related to the former two kingdoms, but they sit right at the boundaries – which is to say, the Mycetozoa became a distinct group around the same time in evolutionary history that animals, fungi and plants went their separate ways.

This is what makes slime moulds in fact deeply interesting. They offer a glimpse of what might have gone on when life began to get truly complex: when single-celled organisms evolved into multicelled ones. In other words, when cell communities started to become superorganisms like us.

Amoebae played a significant role in the history of how we came to understand living matter. The word was coined originally to denote any microscopic organism that doesn't have a fixed shape. Bacteria do: typically they are cigar-shaped, like round-ended tubes. But amoebae are shape-shifting blobs that move by extending a part of their bodies into pseudopods ("false feet"). The term "amoeboid" has entered everyday speech to denote that kind of amorphous, oozy mass.

The amoeba Proteus: a cell of no fixed shape.

But amoebae aren't really a well-defined class of organism at all. There are types of amoebae that are truly animals, or fungi, or plants, as well as protozoans, which are single-celled organisms more "complicated" than bacteria. (I'll say shortly what I mean by that.) Some amoebae are parasites; some are slime moulds. Even some of our own cells display amoeboid behaviour, such as white blood cells that "eat" bacteria and other pathogens by engulfing and absorbing them.

Amoebae were first reported in the eighteenth century in studies of seawater under the microscope. In 1841, the French biologist Félix Dujardin christened the jelly-like contents of amoebae "sarcode". Renamed as "protoplasm", this stuff became regarded as the fundamental living material. Amoebae came to be seen as the exemplar of the living cell, and to some scientists of the late nineteenth century it seemed that complex organisms like us were little more than sophisticated versions of their colonies: English physiologist Michael Foster wrote in 1880 that "The higher animals, we learn from morphological studies, may be regarded as groups of amoebae peculiarly associated together." The German biologist Ernst Haeckel, a Darwinian committed to finding similarities and analogies among living things, attested that the amoeba was a sort of egg cell that needed nothing but itself to multiply: a "permanent ovum", as he put it.

That was the heyday of the amoeba, which by and by came to be seen as too primitive a creature to take us very far in understanding life in all its variety. But if you suspected that amoebae don't seem likely to hold much of interest in their gelatinous lives, *Dictyostelium discoideum* will set you right. It lives in soils and consumes bacteria, helping to maintain a balance in the microbial ecosystem that is as important for the health of the soil as harmony among our gut microbiota is for our own well-being. It's appropriate, then, that *Dictyostelium discoideum* – which I shall call Dicty – was discovered by the son of a farmer, working during the years of the Great Depression when soils were under threat from drought and wind erosion on the American prairies. That man was Harvard microbiologist Kenneth Raper.

What fascinated Raper was that Dicty has a peculiar life cycle. When food or moisture becomes scarce, the cells give up their individuality and turn into a multi-cellular superorganism. They send out chemical signals that attract one another, and the amoeboid cells gather into a "slug" a few millimetres long that contains hundreds of thousands of them. The slug undergoes some shape changes before narrowing at one end and ballooning at the other, becoming a tiny plant-like structure standing upright on a stalk. The bulbous head is the "fruiting body", filled with cells that have become robust spores in suspended animation, ready to be released when conditions are conducive to start the cycle again. In the fruiting body, cells that were once identical have become distinct: they have differentiated, acquiring specialized skills.

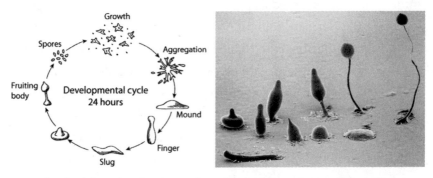

Left: The life cycle of Dictyostelium discoideum. *Some of these forms are shown in sequence under the microscope on the right.*

There is sacrifice involved. The spores will survive, but the supporting tissue of the fruiting body will die. That seemed curious to Raper: these autonomous cells make a choice, some voluntarily renouncing immortality for the sake of the others. It's not unlike the way, during the development of the human embryo, a ball of identical cells apportions into tissues with separate fates. Some become body (somatic) cells, which will die with the person. Others become germ cells, which can in principle keep propagating forever.

What's more, just as this cooperative behaviour of our cells depends on their exchanging chemical signals that allow them to self-organize into pattern and shape, so we see that too in Dicty. There the patterns are remarkable, even beautiful, and certainly adequate to make the case that slime moulds deserve a bit more respect. Some of the cells in the community become pacemakers, exuding pulses of a chemical that diffuses out into the surroundings and induces neighbouring cells to start moving, pseudopod by pseudopod, towards the signalling cell. Because the attractive chemical comes in pulses, the cells advance in waves, resembling the concentric patterns of ripples on water. Eventually these motions coalesce into streams that converge on the place where the fruiting body will grow.

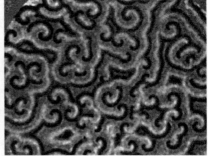

The patterns formed by Dicty cells as they aggregate into multi-celled fruiting bodies.

This behaviour supplies a model system for understanding the appearance of pattern in cell biology more generally. It's not really what human cells do, but there are resemblances. The way Dicty's signalling molecules travel in waves through the cell community is also mathematically analogous to how waves of electrical excitation pass through the cells of the human heart, inducing a steady heartbeat.

Still, Dicty seems deeply alien. Blurring our categories even more, the cells sometimes reproduce by simple division, like bacteria, but sometimes by sex between two of the three different "mating types" – three different genders, if you will.

But as I watched my skin cells turn into a mini-brain in a dish, I had to wonder if we are really so unlike Dicty after all. We are single, autonomous beings, but we are also aggregates of micro-scopic entities that might *each one* give rise to another entire organism. Here were my swarming cells, making their individual ways in the world. They may divide and proliferate, they may cluster into clumps from which organs will grow. They're a part of me, but they can live apart too.

We are not, though, superorganisms in quite the way that Dicty is. For one thing, any pieces of us that become detached will normally perish fast, whereas if you cut off a piece of Dicty's fruiting body it will grow into another fruiting body. Our own cells need to stay and work together for our entire life cycle, whereas when the Dicty spores are revived, they can grow into communities in which single cells can do their own thing again. For Dicty, multi-celled existence is just a passing phase.

Yet the origin of multi-cellularity must have looked a little like this: single cells finding out the benefits of forming temporary unions, of taking specialized tasks, of reproducing sexually. That history used to seem so distant from us – perhaps a billion years ago – that it barely seemed part of our human heritage at all. Now we can see under the microscope that this past has never quite gone away.

As much as a recent shared ancestry with simian cousins, the origin of humans as colonies of cooperating cells was what seemed so unsettling about Charles Darwin's evolutionary theory, which implied a chain of being extending all the way back to amoebic "protoplasmic slime". That we possessed ape-like ancestors might have been deemed undignified. But to collapse the human body to the cell and turn identity into unstructured living matter – that seemed an absurd affront. To some people, it still does.

* * *

Slime moulds are one of the simplest members of the domain of living organisms called eukaryotes, which also includes plants, fungi and animals. What else does that leave in nature? Just single-celled organisms: bacteria and archaea, the so-called prokaryotes.

It has taken much of the century and a half since Darwin to shake off the notion that these distinctions imply a hierarchy of status, with evolution being a progressive elaboration and improvement of living matter at the pinnacle of which is . . . guess who? The simplest way to dispel that illusion is to recognize that all the other types of organism are still with us, many of them thriving (if we let them). Cell for cell, bacteria outnumber us humans by a factor of several tens of millions. So who is truly the most successful?

The question is then why bacteria and other prokaryotes have stayed resolutely single-celled, while many eukaryotes are multi-cellular organisms.

Being a eukaryote is a necessary but not sufficient criterion for getting multi-cellular. The word comes from the Greek for "true/good kernel". It reflects the fact that eukaryotic cells *have* a kind of kernel, while prokaryotes don't – namely the dense cell nucleus where the gene-laden chromosomes reside. Prokaryotes have genes too, but they are not sequestered in a separate cell compartment, and neither are the genes apportioned between several chromosomes as they are in eukaryotes. The genes of bacteria are mostly housed on one double-helical loop of DNA, wound into coils and floating freely in the cytoplasm, sometimes along with several smaller, circular segments of DNA called plasmids.

The organization of chromosomes is just one respect in which the structure of eukaryotic cells is more complex than that of prokaryotes. Along with the nucleus, eukaryotes generally also contain a host of other compartments or "organelles", bounded by membranes, that carry out particular functions: the mitochondrion, the chloroplast, the endoplasmic reticulum and so on. We know what roles these organelles fulfil, but there's a puzzle about that "good kernel", the nucleus itself.

The usual story is that it protects the DNA. But as biochemist Nick Lane asks, protects it from what? What is there to fear in the rest of the cell?

Well, it could be from viruses. But another hypothesis has been suggested by evolutionary biologists Eugene Koonin and Bill Martin: the nucleus is there to slow down the process of protein production from the genome. Recall that the genome of eukaryotic cells (but not that of prokaryotes) is full of rogue bits of DNA called introns that interrupt the gene sequences encoding proteins. It's thought that these introns might be the remnants of an infestation of so-called "jumping genes" – pieces of DNA that are adept at splicing themselves at random places in the genome. Many eukaryotic introns are ancient: they appear in the same places in equivalent genes in a variety of eukaryotic organisms ranging from humans to yeast. This suggests that there was an episode far in the evolutionary past when the genomes of eukaryotes became particularly vulnerable to infestation by jumping genes.*

Whatever the reason, introns now need to be cut out before

---

* All this makes introns sound like a bad thing. But if they were that bad, you'd expect that evolution would have found a way to get rid of them by now. However, the need to edit out introns and splice together the remaining RNA fragments (exons), even though it requires energy to drive the corresponding enzymes, seems to confer benefits. In particular, it creates the possibility of putting together the exons in new permutations, so that a single gene can give rise to more than one protein. This creates more possibility for finding proteins with useful functions. From the 23,000 or so genes in the human genome, around 60,000 different

proteins are made. This happens after the transcription of DNA into RNA, the intermediary molecules that serve as the templates for protein synthesis on the structure called the ribosome. An RNA transcript of a gene is made from DNA, and then it is edited by special enzymes before being used by the ribosome to guide protein synthesis.

In bacteria, this sequence of transcription to ribosomal RNA followed by translation to proteins happens all at once; the RNA is translated even while it is being transcribed. If that happened in eukaryotes, there would be no time for proper intron editing. But the nucleus separates the process of transcription, which happens inside its membrane, from translation, which happens outside. Maybe, say Koonin and Martin, this spatial separation of transcription from translation ensures that the job is done properly.

* * *

Since eukaryotes are more complex than prokaryotes, it seems natural to suppose that the latter cells appeared first and eukaryotes evolved from them. This is indeed what is suggested by both the fossil record (even single-celled organisms leave fossils of a kind) and studies of DNA, from which we can deduce how the "tree" of evolution branched.* However, the differences between prokaryotes and eukaryotes are rather profound, and it's not obvious how to get from one to the other along the gradual steps that evolution tends to take.

---

proteins are made. Some genes may give rise to tens or even hundreds of different proteins. And beyond offering opportunities for shuffling protein structures, some intron fragments themselves have been found to have biological functions – for example, they may help to control the growth rate of yeast and to boost yeast's resistance to starvation.

* The old image of a tree, suggested by Darwin himself, is now understood to be rather more complex – closer to a bush, but also permitting some exchange of genetic material between different branches: the process called horizontal gene transfer that bacteria are adept at.

It's now believed that this isn't how it happened. Rather, eukaryotes are thought to have appeared by the abrupt merging of simpler cells.

The Earth is about 4.6 billion years old, and life seems to have begun at least by 3.8 billion years ago. It consisted of nothing but single-celled prokaryotes for perhaps as much as three billion years after that; the first multi-celled eukaryotes don't appear in the fossil record until around 600 million years ago. No one knows what those first organisms were like, but it's possible that they resembled the slug-like aggregates of Dicty, now permanently united into a single body. Alternatively, they might have been similar to some of today's sponges.* At any rate, they were preceded by single-celled eukaryotes, comparable to the organisms called protists today, which include algae and some amoebae (like Dicty). Multi-cellularity evolved independently many times among different types of eukaryote, which is tantamount to saying that it is a pretty good adaptive strategy in many circumstances.

The bigger question is how eukaryotes arose in the first place. They form one of the three fundamental *domains* of living organism. Bacteria comprise another domain, and the third is made up from that other type of prokaryote, the archaea. Until about 40 years ago, archaea were thought to be just a subgroup of bacteria, until the microbiologist Carl Woese showed by deducing evolutionary relationships from microbial RNA that they are distinct. These studies implied that the first division into domains separated bacteria from archaea, and that eukaryotes later split from the archaea.

---

* That's not the same as suggesting that we are descended from such sponges. One can probably never too often issue the reminder that there is no organism extant today from which we and other complex creatures were descended. Rather, all living organisms exist at the tips of evolution's branches, and we all at various points share common ancestors in organisms that are now extinct.

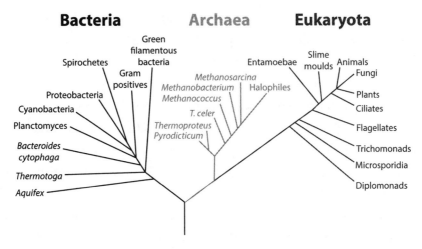

The domains of life. As you see here, multi-celled animals and plants are, from the evolutionary point of view, relatively minor branches of the evolutionary tree (top right), and most life is single-celled. Notice too how "close" to us slime moulds like Dicty are in this representation.

I said that what distinguishes eukaryotes from prokaryotes is that they possess a cell nucleus. That's true, but it doesn't mean that what turned prokaryotes into eukaryotes was the acquisition of such a nucleus. When the "primal eukaryote" – the last common ancestor of all eukaryotes – lived is surprisingly unclear: estimates put it at between 1 and 1.9 billion years ago. But it's generally thought that this organism had many of the key features of eukaryotes today, such as the major cell compartments and organelles. Perhaps the most important of these was not the nucleus, but the energy-producing compartments called the mitochondria – which in that primeval organism were not mere organelles but equal partners in a cellular union.

In the 1960s, microbiologist Lynn Margulis proposed that mitochrondria are the remnants of what were once separate prokaryotic organisms in their own right, which had become "swallowed" by other cells to form a symbiotic relationship. That eukaryotic organelles in general might have originated in such symbiotic mergers was an old idea, proposed in the early twentieth century, but Margulis championed that view in the face of much opposition and ridicule, and it is now accepted.

The question of what swallowed what is still disputed, though. Did two single cells merge like soap bubbles (as some bacteria have been seen to do), or did a type of prokaryote evolve into a cell-swallower (a so-called phagocyte, rather like amoebae and certain white blood cells)? Regardless of how it happened, the acquisition of mitochondria profoundly changed the capabilities of cells, because it supplied them with a new, enhanced source of energy (as well as other new ingredients or abilities). As we all know, when we're more energetic we can do a great deal more – such as work in cooperative teams and communities. Evolutionary change may allow an organism to spread into new "niches": new parts of the environment where there is less competition for resources. For the first eukaryotes, the addition of organelles conferred new functions and literally opened up new horizons.

That is evident too in another cell merger with profound evolutionary consequences. Plants and green algae can perform photosynthesis – the conversion of sunlight into metabolic energy – because they contain organelles called chloroplasts that have the necessary light-absorbing pigments and protein machinery. These capture light and use the energy to pump hydrogen ions across membranes, creating a build-up of charge that, like water behind a dam wall, can be tapped to extract the stored energy. Some bacteria carry out photosynthesis too, and it's generally thought that the common eukaryotic ancestor of plants and green algae acquired a chloroplast by merging with a photosynthesizing prokaryote.

These mergers show how adaptive and plastic cells are. Higher animals often exist in symbiotic relationships, but we don't see them blend into single organisms. Urchin crabs give venomous fire sea urchins a free ride because the sea urchins help them hide from and deter predators. But the two creatures aren't going to become a unified crab-urchin. It's not easy for cells either, but it's possible – because, you might say, their fundamental processes of replication and metabolism are more closely aligned. We're still figuring out how important such mergers were in evolution: microbiologist James Shapiro says that "our understanding of how powerful an evolutionary force cell fusion has been is in its infancy".

And what is sexual reproduction, if not a merging of cells – a union of gametes? This too was another evolutionary innovation that opened up fresh possibilities. It means that the two united cells can swap genes directly: it is really a variant of the horizontal gene transfer that bacteria regularly conduct. It allows for a much more dramatic and rapid reshuffling of a genome than can be obtained from the gradual mutation that happens between one bacterial generation and the next. That is the whole point: sex is another way to evolve. As we saw, it's one option among many; indeed it is many options among many others, since the varieties of sexual reproduction are impressive, especially among fungi.

Notice how an observation about form prompts a question about function. Eukaryotic cells are evidently more complex, with their organelles and nuclei and segmented genomes and so forth, than prokaryotes. But prokaryotes function perfectly well, and so the instinct of the biologist is to ask what eukaryotes gain from their added complexity, especially given that it must be more costly in energy and materials to construct and maintain these internal structures. Evolution tends not to make changes without a reason,* and the reason is usually that the change offers some adaptive

---

* This is a generalization. Evolution relies on random mutations, and so changes appear (mostly) at random. The question is really why particular changes persist. Generally that is because of the evolutionary advantage they convey, but it is possible also for change and variety to appear simply because there is no good

advantage: it improves the prospects for survival and reproduction of the organisms that possess it, by giving them access to a new evolutionary niche. That is what natural selection is about.

It was, then, the appearance of eukaryotic cells on Earth that made multi-cellular existence a possibility. As Dicty demonstrates, this needn't be an all-or-nothing affair. Multi-cellularity can improve survival prospects in some circumstances, compared to going it alone, but this doesn't means that being multi-cellular is universally "better" or makes cells "more likely to survive". Many eukaryotes are doing fine without it.

We many-celled humans evolved not because becoming unified communities of many cells allowed us greater complexity (who cares about that, say 5 million trillion trillion bacteria?), and certainly not because being multi-celled set us on the path to our present alleged cognitive sophistication (those bacteria again: "like we said . . ."). Multi-celled organisms found new niches in which to survive, and to the extent that evolution can be said to have a "goal", it is simply that.

Was multi-cellularity nevertheless an *inevitable* evolutionary development? Now *that's* a tough question. It may not be a truly scientific one, unless perhaps we find some way to rerun evolution among prokaryotes in realistic computer simulations. Maybe it's a rare attribute; perhaps there are countless Earth-like planets in the cosmos with biospheres that have never got beyond the prokaryotic stage even after many billions of years. Or maybe the chances are high enough that the biospheres of most such planets make the transition sooner or later, before their sun is extinguished. This is just one of the many unknowns in the question of how abundant "intelligent" life is likely to be in the cosmos.

---

reason to eliminate it: it conveys neither adaptive benefit nor deficit. Varieties of pigment marking patterns on seashells might be examples of this neutral variation.

CHAPTER 3

IMMORTAL FLESH

HOW TISSUES WERE GROWN
OUTSIDE THE BODY

Even before being subjected to the bio-alchemy that transformed my skin to neurons, it seemed strange and impressive that this piece of me could be kept alive and growing in a petri dish. I'm not surprised that, when living tissue was first grown outside the organism, the feat seemed like an act of occult magic that promised immortality.

It happened at the start of the twentieth century, and even many biologists didn't see it coming. Newspaper reporters forecast an eternal life in which we would be sustained indefinitely by a succession of new body parts, grown in the lab to replace old, worn-out versions. For once the journalists could be forgiven their hyperventilating headlines, because some scientists gave them very good reason to believe such things, speaking of death as a "contingent process" and of cell cultures as "immortal".

It was a classic example of what happens when science runs into mythical territory. Scientists are then as liable as anyone else to adopt fanciful language, allusions and associations. Before you know it, technology is feeding ancient dreams and no one quite knows where the boundary lies between fact and fiction.

This, after all, was the first intimation that our cellular nature disrupts and blurs old categories and certainties, and leaves us reconsidering some of what once seemed the basic facts of human existence. We became less sure of who or what we are.

*　*　*

It was long believed that flesh needs a body to survive. And for good reason: chop off a finger, and that finger is finished. Its cells will stop metabolizing and will decay. It quickly becomes dead flesh: raw material for bacterial metabolism to work on.

It is not the trauma that kills it; people have recovered the use of lost fingers if they are quickly grafted back onto the body. It looks almost as if the finger needs to draw some vital essence from the body to sustain itself. There was a sense, long confirmed by experience in medicine and surgery, that the life of tissues depends on their being a part of the integrity of the body. The individual persists in a way that bits of us cannot. In the late nineteenth century, French physiologist Claude Bernard supposed that the body creates an environment, which he called a *milieu intérieur*, that sustains life. Without this *milieu*, cells must die.

But in 1907, the American embryologist Ross Harrison showed this is not the case – or perhaps that a *milieu intérieur* can be created artificially. He kept tissues alive outside the body in a dish, sustained by a bath of nutrients.

Harrison didn't consider that he had done anything remarkable, for culturing living tissue outside the body wasn't his real goal; it was just a means to an end. He was seeking to settle a long-standing argument about how nerve cells – neurons – grow. Neurons are odd-looking as cells go: they are typically spindly and branching, their filaments looking like microscopic roots.* These filaments are the connective "wires" by means of which neurons join into networks and send electrical impulses to one another. Physiologists in the late nineteenth century weren't entirely sure if neurons were really separate cells at all; the Italian physician Camillo Golgi believed that nervous systems are instead one continuous network. His rival, the Spanish pathologist Santiago Ramón y Cajal, was sure that nerve cells are discrete entities. Even when both Golgi and Cajal were jointly awarded the Nobel prize for physiology in 1906 for their studies of the nervous system, they remained unreconciled; Golgi rather ignobly used his Nobel speech to argue against Cajal's position.

Harrison sought to resolve the matter by growing amphibian nerve cells from pieces of embryonic tissue kept alive in a nutrient medium in jars. Today, it might seem an eminently sensible approach to the question: just look to see if the cells grow as separate entities, like trees from shoots, or as a single unit. But Harrison's experiment was notable not only for the bold idea that tissue might be kept alive outside the embryonic body. It was also an indication of a more general impulse in the study of life at the *fin de siècle*. All these quaintly old-fashioned descriptors we use for the life scientists of that age – physiologist, physician, zoologist, anatomist – reflect the fact that understanding the living world at that time tended to entail careful inspection, observation and categorization, whether it involved whole organisms, their structures and organs, or their microscopic cells. But from around the 1880s, it became increasingly common to intervene in and to manipulate living matter: to prise or shake apart tissues and embryos, and to tamper with their growth. The study of life was

---

* There are in fact hundreds of different types of nerve cell in the brain, each with its own characteristic branching shape – as different as a poplar tree is from an oak.

changing from an observational to a genuinely experimental science: it was becoming *biology*. Harrison's culturing of embryonic amphibian tissue was very much in that spirit. It was still a relatively new and contentious idea that one could learn about living nature by *changing* it. And this is how the centrality of the cell could become recognized. As Hannah Landecker has put it, "In not just taking the animal body apart, but leaving it apart, cellular life that was autonomous, external, and dynamic came into being for biology."

Harrison found that the nerve fibres extend by outgrowth of new neurons: by adding more cells. He also looked at how cells called neuroblasts (a kind of specialized stem cell that can generate several types of neuron) responded to other tissues nearby. Neuroblasts near skin cells would grow into neurons for sensing (for example in touch), whereas neurons near muscle cells grew into the types that drive muscle movements, called motor neurons. This was one of the first reports of how a cell's fate when it differentiates may depend on the nature of the surrounding tissue.

It wasn't immediately obvious that Harrison's experiments, showing that cells can survive and proliferate *in vitro*, revealed anything fundamental about living tissue. Amphibians had long been known to be special – the salamander is unusual among vertebrate animals in its ability to regenerate lost limbs. Besides, it was known already that some organisms can perform much more impressive feats of regeneration. Cut the freshwater polyp called a hydra in half, and each segment will regrow the missing portion. You can even break it down in a blender into more or less its constituent cells and they will reassemble: an echo of our evolutionary history as cooperative cell colonies.

But when the French surgeon Alexis Carrel of the Rockefeller Institute in New York (now Rockefeller University) heard Harrison speak about his work in 1908, he quickly understood that the techniques mattered as much as the conclusions.* Carrel wasn't terribly interested in the argument between Golgi and Cajal, but he was extremely interested in keeping living flesh alive. He had been experimenting with transplantation: removing organs and limbs from animals and then sewing them back on again. His skill at suturing was legendary, and in 1912 he was awarded a Nobel prize for his achievements in joining severed arteries and veins. A major challenge in transplantation was to keep the organ viable while it was removed from the body. Could Harrison's technique assist that?

In the space of just a few months, Carrel and his assistant Montrose Burrows at the Rockefeller worked marvels, adapting Harrison's method until they could culture a wide range of different tissue types: from other mammals (including humans) and birds, from embryos and adult organisms, from healthy and diseased samples. Human tissue proved hard to sustain, generally lasting for just a few days. But Carrel and Burrows struck rich with tissue taken from the heart of embryonic chickens. The added benefit here was that they could tell instantly if the tissue was alive because it pulsed of its own accord, the result of waves of electrical activity that the cells generate and coordinate. That was convenient, but it was also mightily symbolic. What could speak more eloquently of the autonomous life of cells than their ability to produce a "heart-beat" independently of any host body?

Carrel became able to sustain such tissue for ever longer periods. By 1911, he could keep chicken-heart tissue pulsing for weeks, bathing the cells in a nutrient medium made from the ground-up embryonic tissue of dogs and then spun (centrifuged) to extract the vital "juice". It doesn't take much imagination to see how this

---

* It is odd that Harrison never won a Nobel prize for pioneering tissue culture. He was considered many times, but in 1933 the committee considered his work of "rather limited value". To call this a misjudgement – even then – would be too kind.

potion could look like an elixir of life, and Carrel nudged the popular reception of his work in that direction by speaking of death's contingency, calling his chicken-heart culture "immortal".

Despite an appearance of almost monastic intellectual sobriety, Carrel was a showman with a flair for publicity. He encouraged the view that culturing cells was an extremely difficult art, all the better to advertise his own skills. The process was, he claimed, like performing a "delicate surgical operation", requiring "the perfect teamwork of well-trained assistants". To add to his mystical aura, Carrel dressed everyone in his lab, himself included, in hooded black gowns: a piece of pure theatre alleged to be a protective measure against contamination and the perturbing effects of light. (It was true that success depended on ensuring the tissue samples did not get infected by bacteria – there were no antibiotics then to protect against it.) As well as boosting Carrel's reputation for near-miraculous powers, these shenanigans discouraged others from entering the field and gave Carrel less competition. Nonetheless, some other researchers did manage to grow cell cultures, notably at the Strangeways Research Laboratory at the University of Cambridge – but there too the methods took on the status of magic spells that had to be followed to the letter.

He was deep into mythic territory, and he knew it. "Carrel's new miracle points way to avert old age," newspapers announced – "Death perhaps not inevitable." On the twentieth anniversary of Carrel's allegedly immortal chicken-heart culture, the *New York Times* gave as clear an indication as you could imagine of where matters stood. "In the next century, if infection, starvation, physical injury and poison are warded off," it said, the immortal chicken heart "may become as sacred as a venerated religious relic."

To the fury of his superiors at the Rockefeller, Carrel artfully curated the image of a man committed only to his science while at the same time stage-managing his fame. His Nobel prize took this celebrity to a new level, as did his collaboration in the 1930s with the aviator Charles Lindbergh. Arguably then the most famous man in the world after making the first solo transatlantic flight in the *Spirit of St Louis* in 1927, Lindbergh was no scientist. But he was a skilful mechanic, and he designed pumps and other instruments for perfusing organs with blood to keep them viable for transplantation. It was a bizarre, unlikely pairing, but fruitful and sincere. Lindbergh found in Carrel a father figure; the two men shared mutual trust and affection untrammelled by their fame. In 1938 Carrel, in black gown and white skull-cap, appeared with Lindbergh and one of their pumps on the cover of *Time* magazine, above the headline, "They are looking for the fountain of age."

What they were really looking for, though, was a way to preserve what they considered to be the superior civilization of the West. For Carrel was a white supremacist who advocated eugenics as a means of preserving a "superior stock" of humankind. He believed that democracy, a tragic invention of the Enlightenment, was creating a society that preserved the weak, inferior and diseased to the cost of the race as a whole. Despite a Frenchman's instinctive wariness of German militarism, he approved of Hitler's advocacy of racial purity. Lindbergh went further. An earnest and somewhat naïve man, he was seduced by the blandishments of the Luftwaffe and enthusiastically embraced the Nazi regime when he visited Germany in the 1930s. As tensions in Europe racked up, he implored US President Roosevelt to stay out of the hostilities, arguing that Hitler's regime offered the best hope for preserving white Western culture. Carrel was in France when the Nazis invaded, and he accepted the invitation of the Vichy puppet government to establish a so-called Institute of Man to pursue his vision of human "perfectability". He died in 1944 after the Liberation while awaiting trial on accusation of collaboration.

For Carrel and Lindbergh, sustaining life outside the body was part of a broader agenda to preserve culture itself. The subject was soon to become embroiled in racial controversy in other ways, and we will see later that the eugenic associations of cell biology and embryology have never gone away.

Carrel's chicken-heart culture never was immortal. It is not clear quite how he kept it alive for decades (the samples were finally discarded four years after his death), but there seems little question that the original cells could not have proliferated for that long. In the 1960s, cell biologist Leonard Hayflick showed that mammalian cells can only divide for a limited number of times – about 30 to 70 – before they expire in a process of self-inflicted cell death. That's a precaution installed by evolution against the accumulation of cell damage and somatic mutation that might otherwise lead to cancer. Hayflick speculated that Carrel's culture might have been inadvertently replenished either by contamination or by cells remaining in the "embryo juice" in which it was bathed. One can't rule out the possibility too that Carrel indulged in a bit of cheating. For mammals, that's still the only route to immortality.

* * *

In the early days of tissue culture, it wasn't just the press and the public but biologists too who struggled to assimilate the implications. An immortal heart in a jar sounded like something out of a Gothic fantasy, and indeed the *Indianapolis News* said of Carrel's work that it had "the creeping horror of the most morbid narrative of Edgar Allan Poe, with the additional shiver that it is the truth and not the product of a fantastic imagination." Thomas Edison, never one to refrain from mixing his spiritualist inclinations into the latest technology, said of Carrel's organ-preserving experiments:

> If some day the scientists arrive at a point where the human body, after life is extinct, can be thus preserved and after an indefinite

time, through the transfusion of life-giving blood or fluid be brought back to resume its normal functions, who can say that we may not learn definitely that there is consciousness after death?

Resurrection of the body is of course an old myth. But tissue culture offered a vision of it that no one could have previously imagined: immortality of the flesh. That point was made by the physician Thomas Strangeways, the first head of the Cambridge Research Hospital that was soon to bear his name. The hospital was established in 1905 as a privately funded institution that admitted patients suffering from chronic diseases, which its staff would research. But from around 1919, Strangeways became captivated by tissue culture as a method to study rheumatoid arthritis via the *in vitro* growth of chicken cartilage. Soon the hospital stopped taking patients and became purely a research laboratory dedicated to understanding the living cell – which Strangeways regarded as an autonomous living entity.

In a 1926 lecture called "Death and Immortality", Strangeways gave his audience an image entirely worthy of the grisliest of Poe's tales. He asked them to imagine a dead body being minced up and made into sausages. No one would doubt that the person was then well and truly dead. But provided that the sausages were kept in cold storage, said Strangeways, one might, days or weeks later, pull out a piece of the meat and grow it into a culture of living cells.

To prove the point, Strangeways there and then produced a culture he had made from a sausage. He said he had got the meat from his local butcher, for which we must take his word. At any rate, the sausage was in some respect still alive.

Poe might have scripted the end of this tale too. After cycling home from that lecture, Strangeways suffered a brain haemorrhage from which he never recovered consciousness. It might seem callous to imagine he would have liked his colleagues to put his thought experiment to the test, but it is hard to dismiss the idea.*

The work at the Strangeways laboratory on tissue culture was cited in *The Science of Life* (1929) by biologist Julian Huxley and his co-authors H. G. Wells and Wells's son George Philip Wells as an example of how biology was "bringing life under control". In an echo of the sausage tale, they wrote that if Strangeways himself had been alive in the time of Julius Caesar, "fragments of that eminent personage might, for all we know to the contrary, be living now."

The mutability of tissues and flesh was explored by Wells in perhaps his darkest and most vivid book *The Island of Doctor Moreau* (1896). Its narrator Prendick is marooned on an island where Moreau, a stranded and deranged British surgeon, remodels animals by vivisection to render them human-like. Wells's book was evidently the template for a short story by Julian Huxley called "The Tissue-Culture King", the sole example of his literary talents (which were not, sadly, to quite the same standard as those of his brother Aldous†). The tale made a sober debut in the scholarly *Yale Review* in 1926 before being brought to a wider audience the following year in the pulp science-fiction magazine *Amazing Stories*, a reliable barometer of the cultural reception of new science and technology.

In "The Tissue-Culture King", a piece of flesh becomes just the kind of holy relic that the *New York Times* imagined of Carrel's immortal chicken heart. The story tells of a remote African tribe whose king Mgobe has his flesh grown to make objects of worship

---

* I can't forego piling on the grotesquerie by pointing out that Strangeways's middle name was Pigg.

† The island trope used by Wells and Huxley goes back to *Gulliver's Travels* and *Robinson Crusoe*, if not indeed to Thomas More's *Utopia*, as a vehicle for imaginative exploration unfettered by the normal rules of civilization. Aldous Huxley used it for his final novel *Island* (1962), which describes a utopian society where the sexual repression and taboos that he regarded as so corrosive in the West have been overcome.

by another rogue biologist named Hascombe. Like Moreau, Hascombe takes advantage of his remote situation to pursue experiments that would be forbidden or rejected among scientific peers.

The narrator is captured by the tribe – which includes physically anomalous individuals such as eight-foot giants and squat dwarfs – and taken to their town, where he discovers all manner of physical peculiarities among the inhabitants: transformations wrought by Hascombe to demonstrate his mastery over human flesh.

Captured 15 years previously, Hascombe saved his skin by showing the tribe a "great magic". He demonstrated the microscope he had brought with him, revealing how human blood is composed of individual corpuscles. Then he persuaded the tribespeople to help him set up a primitive laboratory that he turned into a "Factory of Majesty", or as Hascombe himself thinks of it, an Institute of Religious Tissue Culture. "My mind went back to a day in 1918 when I had been taken by a biological friend in New York to see the famous Rockefeller Institute," says the narrator, "and at the word tissue culture I saw again before me Dr Alexis Carrel and troops of white-garbed American girls making cultures, sterilizing, microscopizing, incubating and the rest of it."

Recognizing the sacred authority placed in the king's physical personage, Hascombe has persuaded Mgobe to allow him to take "small portions of His Majesty's subcutaneous connective tissue under a local anaesthetic". He has cultured this into pieces of living flesh, which are dispensed to the subjects as items of supernatural power. As additional fetish objects, he has bred grotesque forms of animals such as the two-headed toads that the narrator sees at the story's start. After making his escape, the narrator concludes with an awkward piece of moralizing – all the more incongruous coming from an author who professed a deep belief in science as a force for social improvement:

I commend to the great public the obvious moral of my story and ask them to think what they propose to do with the power which is gradually being accumulated for them by the labors of those who labor because they like power, or because they want to find the truth about how things work.

Beyond its Eurocentric condescension and prejudice towards other cultures, Huxley's tale thus expresses ambivalence towards the emerging biological technologies that revealed the mutability of living form and matter. It was almost as though fiction could supply a vehicle for thoughts and fears that Huxley the scientist could not quite articulate. In that respect, fiction seemed to serve a similar role for Huxley as it did for Wells, who in his non-fiction came closer to boosterish advocacy of science; he furiously denounced Aldous Huxley's own take on the prospects of growing humans in *Brave New World* as a betrayal of the future.

* * *

After their founder's death in 1926, researchers at the Strangeways Laboratory stepped up efforts to popularize the science of tissue culture. Thomas Strangeways's successor was a young zoologist named Honor Fell, who gave public talks and wrote about the research there. The work was also vividly advertised in time-lapse films of cells moving and proliferating under the microscope, made by the pathologist Ronald Canti. These movies were reviewed in the *Times* and even screened privately for the prime minister Ramsay MacDonald and the Duke of York. Blown up to the size of animals, the amoeba-like blobs took on a life of their own.

In 1930, Fell spread the word to the nation in a BBC radio broadcast titled "The Life of a Cell" – implying that the cell should be regarded as an independent organism. Fell suggested that, far from creating some artificial form of life, tissue culture let us see the "real" cell set free from the complicating influence of the body.

But was that freedom really to be desired? Biologists had come to appreciate that cancerous tumours were caused by cells losing their customary discipline and multiplying with abandon. Some biologists worried that cells in culture were more like this than like healthy tissue. In *The Science of Life*, Huxley and the Wellses likened cells in organs to people in "the City [of London] during working hours", while cells in culture looked more like "Regent's Park on a bank holiday, a spectacle of rather futile freedom". That imagery may have been rather benign and carefree, but underneath it runs a fear of disintegration of the social order, a reversion to a primitive "state of nature" – even (like Moreau's Beast-Men) to a more primal stage of the evolutionary scale. There are strong echoes here of Virchow's politicized cell theory, albeit with no enthusiasm for the anti-authoritarian anarchy that he seemed to welcome.

It was no surprise, then, that in popular media the optimism of the Strangeways view of cell culture curdled into something darker. Preceding Huxley's tale in an earlier issue of *Amazing Stories* that same year was a story called "The Malignant Entity", in which a cell culture goes wild and devours its creator and several others before being killed with poison.

Flesh living outside of, or *beyond*, the body is a stock trope of creepy tales. The severed hand crawling towards its victim had become camp well before a young Oliver Stone rather unwisely elected to make the B-movie *The Hand* (1981), while Poe's "The Tell-Tale Heart" (1843) conjures up the heart of a murdered man beating after death when the murderer hides the dismembered corpse under the floorboards. Zombie narratives often attribute an independent animation to every limb and lump of the "living dead", and this notion too owes something to Carrel's work on sustaining flesh and prolonging life.

Carrel's research was again blended with the Moreau archetype in a 1927 story in *Amazing Stories* which anticipates the zombie genre from its title – "The Plague of the Living Dead" – onwards. Once again an outcast biologist, here called Farnham, carries out his dodgy experiments in some distant land where the prejudices of the day permit the author to spin fantasies of savagery and conquest among other races. On a Caribbean island, Farnham tries out a serum of reanimation on inhabitants killed by a volcanic eruption, only to end up pursued by a mob of the undead: "Doctor Farnham had a fleeting, instantaneous vision of the two fellows being chopped into bits or torn to pieces and each separate figment of their anatomies continuing to live." Four decades later, George Romero appropriated this imagery in his smartly political low-budget shocker, which turned the prejudices of those old stories on their head. The hero of *Night of the Living Dead* is a black man, shot dead in the end by a posse of white men who mistake him for a zombie while all too clearly representing a modern lynch mob.

Cultural and feminist theorist Susan Merrill Squier astutely recommends that we resist the temptation of a simple dualism that makes literature "the unconscious of science". Nonetheless, it is clear from Huxley's Tissue-Culture King that literary modes of expression can refract the social meanings of science in valuable ways. Researchers involved with the Strangeways Hospital supplied another instance of that when they indulged an urge to contextualize their work on cells through poetry. Conrad Hal Waddington, who visited there in the 1930s, expressed his thoughts in stanzas depicting the operation of "organizers" that direct embryo development (see page 66) as a kind of dance between the autonomy of the individual and the inclinations of the collective:

> Now every separate part is tied
> to particular performance
> but still within itself is free
> to organize its own affair.

Meanwhile the description by Strangeways researcher Arthur Hughes of the tissue culture of mammalian sex organs by his colleague Petar Martinovitch seems to echo the opening scene of *Brave New World*:

> Testicles and ovaries
> Explanted in a row
> Grown by Martinovitch
> In vitro.

Martinovitch's reply to his colleague's versification shows how the researchers liked to adopt a cell's-eye view of their work, anthropomorphizing them and making the cell or the tissue culture a stand-in for the whole organism, with every bit as much autonomy and agency:

> At the Strangeways lab a certain lad
> Sits on the bank (always left) of a blood vessel and follows;
> Myriads of little creatures
> Of similar birth but different features
> Carried by a stream of swift motion,
> With powerful sweep and great commotion –
> To their unknown destiny.

It's not clear what impulse led the Strangeways scientists to this decidedly unusual way of processing their work, but the whimsical cell's-eye view was adopted too by Honor Fell. "There is something rather romantic," she said, "about the idea of taking living cells out of the body and watching them living and moving in a glass vessel, like a boy watching captive tadpoles in a jar."

* * *

Carrel helped to make tissue culture a general procedure for all kinds of cells, including those of humans. But despite the headlines proclaiming imminent immortality, cultured human tissues were hard to keep alive.

That changed in 1951. When the American doctor George Gey at Johns Hopkins Hospital in Baltimore removed cancer cells from a 31-year-old patient named Henrietta Lacks, he found that they were like no other sample he had seen. Lacks died of the cervical cancer later that year, but her cancer cells continued to proliferate in culture, seemingly without limit. The extraordinary vitality of these so-called HeLa cells (thus identified by the conventional anonymized abbreviation of the donor's name) soon made them the standard strain for human cell experiments worldwide. In particular, they were used for testing drugs without placing living people at risk. Gey began using HeLa cells as a host tissue to grow viruses, which cannot persist without cells to colonize. By 1954, such work had yielded a vaccine for the polio virus, discovered by the biologist Jonas Salk.

The HeLa story has been told many times in newspaper articles, television documentaries and books, the most powerful and comprehensive being Rebecca Skloot's *The Immortal Life of Henrietta Lacks*. As Skloot's title implies, the mythical tropes have become so thoroughly mixed into this narrative that it is hard to find a vantage point that reveals the full context. Hannah Landecker identifies what is at stake: these accounts, she says, "are responses to something otherwise not easily comprehended in narrative: infrastructural change in the conditions of possibility for human life". HeLa, says Landecker, was "living proof of the unexpected autonomy and plasticity of the human somatic cell".

The story has taken the shape of a parable. It appears to speak of the mistreatment of minorities and disadvantaged social groups in medicine, in an age before proper ethical regulation. That Lacks's cells were kept and used by Gey without her consent was normal practice for the time, but the story highlights the gulf that separated a poor black family from the world of modern medicine and research in the 1950s. The Baltimore hospital was already widely feared by the black community, wherein rumours circulated that black patients were experimented on in the basement. This was not mere paranoia born of racial tension and mistrust on the faultline between the American north and south. No secret experiments were conducted at Johns Hopkins, but in 1932 black sharecroppers had been recruited at Tuskegee University in Alabama for a study of the progress of untreated syphilis. The volunteers were given free meals and health care, but some were infected with the disease and none was given treatment with the antibiotic penicillin, known to be effective against it. The study continued for an astonishing 40 years until a whistleblower revealed what was happening.*

But the story of Henrietta Lacks is more complicated than a tale of exploitation. It speaks partly to the distance between biomedical research and public understanding of it and its motivations at that time. Lacks's family was baffled and disturbed for many years about what had happened to her. They suspected that the doctors had stolen her cells and profited from them, and had denied her relatives any share of those profits. They were understandably confused and angry. Yet Gey himself never sought to make money from these cells, which he delivered, sometimes by hand – kept warm in a test-tube in his shirt pocket – to colleagues who asked for them. Traffic in human cells and tissues from the operating theatre to the research laboratory was considered perfectly acceptable at the time and conducted in an open manner. Neither scientists nor the public seemed to feel there was anything untoward about it.

---

* After the discovery of HeLa cells, a laboratory was set up to mass-produce them. It was based at Tuskegee.

The case of Henrietta Lacks also highlights the confusion of categories that tissue culture created. Without any knowledge of cell biology, and never having had the process explained to them, her family was unclear whether Henrietta herself was not somehow being kept alive, perhaps even still suffering. It is tempting to attribute these misunderstandings and fears to the Lackses' paucity of formal education and their suspicion of authority – a matter that could have been resolved if they'd had access to the right information. And indeed some of that anxiety was laid to rest for Lacks's daughter when, sensitively accompanied by the intrepid Skloot, she was taken into a Johns Hopkins lab in 2001 and shown HeLa cells by a young researcher. But the Lackses' responses were, at root, entirely apt – for *no one* knows quite how to think about the "immortal life of Henrietta Lacks".

Consider, for example, the description of the events given in a 1971 tribute to Gey after his death, in the journal *Obstetrics and Gynecology*. The tissue sample that he took, say the authors,

has secured for the patient, Henrietta Lacks as HeLa, an immortality which has now reached 20 years. Will she live forever if nurtured by the hands of future workers? Even now Henrietta Lacks, first as Henrietta and then as HeLa, has a combined age of 51 years.

What is going on with this strange identification by these academic authors of the cells with the person? They are no more able to frame that state of affairs than the Lacks family were. What's more, the authors offer up the same kind of narrative of an out-of-control cell mass as we saw in the gruesome tales of *Amazing Stories*, writing that, "If allowed to grow uninhibited under optimal cultural [!] conditions, [HeLa] would have taken over the world by this time."

Taken over the world. In these retellings of Henrietta Lacks's story there is, Landecker says, "a constant preoccupation with mass

– what she would weigh now." The figure – the mass of HeLa cells cultivated from those taken from Henrietta's tumour – was estimated by one scientist in the early 2000s at 50 million tons. Such crazy, meaningless figures reveal – or is it conceal? – an attempt to grapple with the notion of life that overflows its mortal container.

We are, albeit less decorously, back with Carrel's monstrous chicken towering over the land. Yet now this outsized, threatening being is not a fowl but a black woman, and in its shadow lurk all the racial anxieties that bedevil the United States.

It is only the flipside of this image of monstrosity that transforms Lacks into an angelic figure granted a sort of eternal life for the benefit of all humankind. This was how she was portrayed in a 1954 article in *Collier's* magazine on HeLa cells, as a Baltimore housewife "thrust into a kind of eternal life of which such a woman would never dream," in Landecker's words – with no mention of the fact that she was poor and black. It's not obvious that we do better now to turn her into the powerless victim exploited because of her race, or indeed into a kind of latter-day saint whose mortal remains proliferate beyond plausibility as sacred relics.

Race is a constant and inevitable theme in the HeLa story. In the late 1960s, geneticist Stanley Gartler claimed that these cells were highly invasive, to the point where many other cell lines used in research, both human and from other animals, had been colonized by them. He sought to identify HeLa cells by looking for the presence of a biological "marker" – a genetic variant of an enzyme involved in red blood cell metabolism – that was specific to those cells. This marker, Gartler pointed out, had been found only in African-Americans.

It was, in fact, only at this point that the race of the HeLa donor, previously undisclosed, became an issue. HeLa cells became "black" and "female", and they were "aggressively" and unstoppably contaminating other cell lines – mostly taken from white folks. As some researchers wrote, it took only a single stray HeLa cell to "doom" a culture.

Under cover of a neutral reporting of "biological facts", what perilous stories science sometimes tells.

You might be wondering why HeLa cells are so useful at all, given that they are not even regular human cells but abnormal, cancerous ones. But for the purposes to which they are put, this does not really matter. As a host for viruses – to explore vaccines, say, or for the research conducted since the 1980s on HIV and AIDS – cancer cells work perfectly well. Likewise, potential drugs may be as toxic to HeLa as they would be to ordinary human tissue.

All the same, there clearly *is* something that sets HeLa apart from other cells, making them such a vigorous and robust strain. It seems to have something to do with their telomeres: segments of DNA at the ends of the chromosomes that are degraded and shortened on each cycle of replication (see page 133). Cancer cells in general produce an enzyme that repairs telomeres and so protects against this form of cell ageing – but HeLa cells appear to be especially adept at this.

One thing is clear: HeLa is no longer simply a specific and unique kind of cell. After this much proliferation, it is inevitable that the cell lines will have accumulated many mutations: they have evolved, subject to all kinds of selective processes. In fact, evolutionary biologist Leigh Van Valen has argued that HeLa should be regarded today as a separate species, distinct from humans, which he called *Helacyton gartleri** – a kind of microbe bred by humans from humans. By no means all biologists accept that idea, but it offers a striking reminder of the blurred boundaries between organism and cell community. In Van Valen's picture, HeLa has been forced,

---

* The name, as you doubtless guessed, makes reference to Stanley Gartler and his studies of the "invasive" nature of HeLa.

during decades of *in vitro* culture, along a sort of reversed evolutionary trajectory to a "simpler" state of being.

But surely HeLa still has human DNA – indeed, Henrietta Lacks's DNA? Well, not exactly. As with bacterial strains, the cells evolve and adapt to prevailing conditions, and have developed distinct forms with quite different nutritional needs and metabolic processes – different "individuals", you might say. In 2013, researchers in Europe sequenced the genome of one strain of HeLa cells and found, unsurprisingly, that it was a mess. Many cells have extra copies of some chromosomes, and many genes have acquired extra copies or have been extensively reshuffled. Some of these changes might have been present in, and responsible for, the original tumour, but it seems likely that many have appeared subsequently. At any rate, the findings raise questions about how good a proxy HeLa really is for the human body, let alone how much we can continue to assert that these cells represent an "immortal" woman who died over half a century ago. What is perhaps surprising is that the accumulation of chromosomal abnormalities – which would make it impossible, say, to use HeLa to clone a human being even in principle – seems to do nothing to diminish the vitality of the cells themselves. It's likely that this resilience to chromosome defects reflects the fact that HeLa cells only need to use a small portion of their genome: all they need to do is keep dividing.

* * *

Tissue culture complicates, as no other biological advance has done, the connection between the life of the individual and the life of their cells: between flesh and body. That complication demands new narratives, a task that Susan Merrill Squier has aptly described as an attempt to "replot the human".

Human tissue is treated in laboratories now as if it were a kind of material, like a polymer or ceramic. There is no disrespect in this attitude; indeed, in my experience people who practise tissue culture think (and *must* think) carefully about the ethics of their work. All the same, as Julian Huxley's curious story recognized, tissue is more than this: more than an active, responsive material, a biomedical resource, a convenient host medium for pathogens and test-bed for drugs. As social scientists Catherine Waldby and Robert Mitchell say in their study of organ and tissue transplantation and donation, *Tissue Economies*:

> Tissues that move from bodies to tissue banks to laboratories to other bodies bring with them various ontological values around identity, affective values around kinship, ageing, and death, belief systems and ethical standards, and epistemological values and systems of research prestige as well as use values and exchange values.

Some of those values are economic and legal. Research on tissues can't operate without a legal framework for establishing the boundaries of the permissible. I was taken by surprise (although not in the slightest bit disgruntled) to discover that once my fibroblast cells had divided *in vitro*, they were no longer legally a piece of me but were classified as a cell line. This enables scientists using tissue cultures to assert intellectual property rights over knowledge gained from these living entities bearing the genome of an individual (live or dead). Personally I'd have been delighted if Selina and Chris had found a way to derive knowledge with commercial as well as scientific value from "my" cells. (Sadly, but just as expected, they were never so special.) But this isn't always the way the transaction goes. In 1984, an Alaskan engineer named John Moore brought a court case against his former physician David Golde of the University of California at Los Angeles, after Golde derived a profitable "immortal" cell line from cancer cells surgically removed from Moore's spleen.

These tumour cells turned out to produce a protein that stimulates the body's immune system to fight infection.

The court ruled that Moore had no property rights over the cells, nor any claim to profits derived from them. They were simply "discarded tissue" – a kind of waste.* That decision benefitted potentially life-saving biomedical research, which would be immensely obstructed if all original donors of tissue from which cell lines are generated could assert ownership. It was also reasonable insofar as it recognized that turning Moore's cancerous cells into a viable source of the protein demanded considerable technical ingenuity on Golde's part. But it also reflects a regulatory system that works to transfer rights from donor to recipient. In the UK, for example, IVF clients who agree to the use of their spare embryos in research must make the donation as a "gift" with no commercial strings attached, even though the researchers who use them can claim property rights over any product derived from this tissue.

There is nothing obviously bad about that – as I say, it enables valuable research to happen. But it reminds us that flesh is, among other things, now a commodity.

And haven't we heard that idea before?

> If you repay me not on such a day,
> In such a place, such sum or sums as are
> Express'd in the condition, let the forfeit
> Be nominated for an equal pound

---

* Despite being legally vindicated, Golde seems to have remained deeply troubled by the incident. In 2004, he committed suicide. That the shadow of the court case weighed heavily on him was implied in a statement by the Chair of Medicine at Harvard Medical School: Golde, he said, "deeply respected the rights of patients and their integrity but he also believed that science should be unfettered with regard to what amounted to discarded tissue."

Of your fair flesh, to be cut off and taken
In what part of your body pleaseth me.

When Shylock is told he may take his quota of flesh from Antonio only if it does not shed a drop of the man's blood (indeed, in an escalation of the antisemitic tensions, of "Christian blood"), he is unwittingly given a glimpse of the market forces that would come to play in the "tissue economy": which tissues, how much, who has ownership of what? Richard Titmuss, a pioneer of social policy on public health and welfare, argued in the 1970s that "the human body exists beyond relations of commerce . . . its value is intrinsic and unquantifiable." But that was more an expression of hope. In reality the market in tissues – with its payments for blood donation and underground trade in organs – would fall in line with the usual pattern for any commodity uniformly distributed across the population, whereby the poor end up selling to the rich.

The dilemma created by Shylock's demand exists because it plays out in a zero-sum game: Shylock's gain of a pound of Antonio's flesh must be balanced by Antonio's proportionate loss of the same. Tissue culture changes that game. It makes the generation of our flesh and blood an industrial process of mass production. Alexis Carrel could have offered Antonio a safe solution, although Shylock's actions might well have confirmed his suspicions about the problem with "bad Jews".

The social and philosophical questions raised by tissue culturing are more than operational, however. They are questions about who we are. These are not to be settled by legal diktat. "What does it mean," ask Waldby and Mitchell,

when the human body can be disaggregated into fragments that are derived from a particular person but are, strictly speaking, no longer constitutive of human identity? How is the status of the individual (strictly speaking the in-dividual, he who cannot be subdivided) altered to accommodate these possibilities for fragmentation?

As a part of me – and yet not (legally, at least) a part of me – grew into something resembling a brain in an incubator five miles across town, I asked myself that more than once. I haven't yet found an answer.

* * *

In 2008, the Museum of Modern Art in New York displayed a tiny "leather jacket" made not from cow hide but from mouse tissue: cells derived from mouse embryonic stem cells grown on a polymer scaffold that guided them into the shape of the diminutive garment. The jacket didn't last long. A few weeks into the show, called "Design and the Elastic Mind", a sleeve fell off, while cells also began to separate in clumps from the polymer scaffold. In the end, the show's curator Paola Antonelli was forced to stop the culturing process. "I had to make the decision to kill it," she said:

> And you know what? I felt I could not make that decision. I've always been pro-choice [on abortion rights] and all of a sudden I'm here not sleeping at night about killing a coat . . . That thing was never alive before it was grown.

The artwork, called *Victimless Leather*, was made by "bio-artists" Oron Catts and Ionat Zurr of the SymbioticA laboratory of the University of Western Australia, a unique collaboration of artist-provocateurs working within an academic establishment to challenge – by actually using – biotechnologies such as tissue culture. The "mouse-cell coat" was, the artists say, intended to critique the idea of in vitro production of leather-like materials as a solution to problems associated with the manufacturing of animal leather for human consumer goods, and was part of a larger body of works that dealt with the ways in which technology is being used to obscure, rather than eliminate, the victims of our consumption.

One of SymbioticA's earlier works was *Pig Wings*: three small wing-like structures made from pig bone-marrow stem cells on polymer scaffolds that were intended as a comment on the "genohype" that the artists perceived to surround the Human Genome Project. That the wings looked decidedly underwhelming was the whole point. The work, said Catts and Zurr, adopted a deflationary "aesthetic of disappointment":

> People ... would be drawn to see the piece because they believed that flying pigs and other biotechnological wonders would be presented to them. Instead they would be confronted with tiny, humble-looking detached wings, made of tissue, which will never fly.

Artistic responses like this may supply an arena where our confusions about the technologies of living matter – and how they might be used – can be explored. What happens when such matter becomes a commercial product? What qualifies as life, and what duties do we have towards it? Where are the boundaries, both physical and ethical, of what we might grow? The science informs and constrains these questions but cannot answer them. The art of flesh developed by SymbioticA *reminds us to be unsettled*, and not to normalize what should in fact be destabilizing, exciting, perplexing and disturbing. Tissue culture began a century ago, but we have yet to come to terms with what it means.

SECOND INTERLUDE

HEROES AND VILLAINS

CANCER, IMMUNITY AND OUR
CELLULAR ECOSYSTEM

We populate the world of cells, like the world of people, with heroes and villains. There are cells that are our friends and saviours, and cells that want to kill us. There are "good" and "bad" bacteria. There are killer cells, rogue cells, zombie cells. The metaphor of cell as organism is thriving and filled with personalities.

Pathogenic bacteria – "germs" – were easily assimilated into this menagerie: they are invaders from outside, the invisible enemy with which we are always at war. But cancer cells are harder to conceptualize, for the same reason that they are harder to combat. They are a part of us: our dark side.

Cancer is not a disease like any other. It's not ultimately caused by a pathogen like a virus or bacterium, although some viral infections can trigger it. Indeed, part of the puzzle, as well as part of the reason why "curing" cancer is such a challenge for medical science, is that many causes all result in the same symptom: the uncontrolled proliferation of cells as tumours, which wreak physiological havoc that can be catastrophic.

Cancer is not, as is sometimes implied, a modern affliction. It has always been with us, and it afflicts most other animal species. While it's understandable that we should frame such a potentially lethal condition in terms of dysfunction, breakdown and invasion – a condition on which we must "wage war" – it's not clear that this helps us to comprehend or come to terms with cancer. It makes more sense to see it as *something cells naturally do*: as an inevitable consequence of being multi-celled beings.

One increasingly popular and fruitful way of thinking about cancer is in evolutionary terms. For it highlights a puzzle about our situation as cell communities, which is that the normal evolutionary process has been largely eliminated from it. As bacteria multiply and a colony spreads, random mutations occur during cell division, and this gives natural selection variation on which to work. That's how antibiotic resistance arises: in the presence of antimicrobial agents, bacterial cells that acquire by chance some resistance gain a strong advantage and soon come to dominate the colony. The more you ply bacteria with antibiotics, the stronger becomes the selective pressure that can generate resistance.

You would expect variation also to arise from copying errors in DNA replication as our own cells divide and proliferate during growth and tissue replenishment. And it does: as we saw earlier, the human body is full of small differences in the precise sequence of our genome. Mostly these mutations have no effect, but it's reasonable to imagine that some will convey a reproductive advantage, for example by letting some cells divide faster.

That sort of "selfish" cell proliferation is precisely what gives rise to cancer, so it's important for the reproductive success of the whole organism that it is kept in check. The human body contains several mechanisms to actively control "*in vivo* evolution" that might otherwise get out of hand. For a start, fidelity of DNA replication in cell division is scrupulously checked by "proofreading" enzymes. There are genes that regulate the cell cycle, acting as brakes to keep cell division under control. What's more, our immune system is constantly on the lookout for pathologically dividing cells. That's why cancer may also result from immune suppression.

One theory of cancer even goes so far as to say that it is not a failure of the body's mechanisms for keeping cell proliferation in check, but rather a natural response to environmental factors that subject cells to stress, which is encoded in our genomes from some very ancient evolutionary epoch but normally kept on a leash. "It may be *triggered* by mutations, but its root cause is the self-activation of a very old and deeply embedded toolkit of emergency survival procedures," suggests physicist Paul Davies. That's a controversial and minority view, but it illustrates how cancer may need new narratives in comparison to most stories of disease. From an evolutionary perspective, cancer is just the kind of thing you'd expect cells to do. Just as civilization and socialization require us to suppress ruthless, selfish atavistic urges, so cells in an animal body must learn restraint and cooperation, and be policed to ensure compliance. It's just as Thomas Hobbes said in *Leviathan*: the body, like the body politic of the state, relies on a suppression of the State of Nature. With cancer, as with anarchy, the alternative is a life that is nasty, brutish and short. (Let me just remind you to keep your Metaphor Alert switched on here.)

* * *

Military imagery – a "war on cancer" – adds to the impression that we are fighting off some invading pathogen rather than trying to keep control of our own cells and flesh. That confusion is invited by the fact that things entering our body from outside can indeed trigger cancer. They do so by interfering with cell regulatory mechanisms that normally suppress tumours.

One of the earliest associations of cancer with environmental factors was the discovery in 1775 by English physician Percivall Pott that there was an unusually high incidence of cancer among men who had worked in their youth as chimney sweeps – soot being, we now know, horribly abundant in carcinogenic chemical compounds. The problem for understanding the causal factors was that these associations seemed to be so varied.

Take X-rays: discovered in 1895, they were at first thought to be an amusing and even beneficial form of radiation for several decades, until they were found to be associated with skin cancers and leukaemia. The same with radioactivity: miners of uranium ore (pitchblende) proved to be particularly prone to lung cancer, and both Marie Curie and her daughter Irène Joliot-Curie died prematurely from cancer surely caused by their experiments on radioactive materials. It wasn't until the link between smoking and lung cancer was discovered in the 1950s that these diseases became associated with lifestyle more generally: what you eat, where you live and work. Many chemical agents, some of them common in foodstuffs and additives, are now known to be carcinogenic: liable to create a raised cancer risk. To complicate matters even more, some cancers are linked to viral infections. The sexually transmitted but largely innocuous human papillomavirus, for example, can increase the risk of cancer, especially of the cervix.

It took a long time to figure all this out: to appreciate that these external agents may in some way or another introduce mutations to genes vital for keeping cell division under control. Some of these genes, called proto-oncogenes, perform important roles in the cell cycle, and a mutation – sometimes of just a single base in the DNA sequence – can convert them to a faulty form called an oncogene that triggers uncontrolled division. A gene called Myc is one of these proto-oncogenes, having roles that are many and varied.*

---

* The rather random names that genes acquire tells a story in itself. Myc

Other cancer-causing mutations appear in genes that actively inhibit the formation of cancer cells, for example by slowing down cell division, repairing damaged DNA or telling cells when to die. These are called tumour suppressor genes, and if mutations disrupt their function, cells may again run amok. One such gene, called p53, is a component of the cell cycle's braking machinery: when it is switched on it can either halt the cycle and initiate DNA repair or induce cell death if the DNA is beyond mending. Both the activation of p53 and its knock-on effects are complicated, but about half of all cancers are thought to involve p53 in some fashion.

These cancer-causing gene mutations can be produced in the body by the action of some carcinogenic chemical or radiation (ultraviolet light or X-rays, say), or they can arise spontaneously during cell division, or be inherited. Recent years have seen a debate about whether genetic mutations appear primarily at random or because of lifestyle – an incendiary question because it carries implications about whether cancer is to some extent just "bad luck" or primarily "self-inflicted".

As well as having tumour suppressor genes, our healthy cells

---

was first isolated from a virus that triggers a certain sort of tumour growth called myelocytomatosis in birds. The naming invites the idea that Myc and its oncogenic form do very specific things related to this viral condition. That's far from the case. There is in fact a whole family of Myc genes, and it's not possible to say concisely what they do except that they have an extremely general role as transcription factors that regulate the expression of a whole suite of other genes, some related to cell proliferation. It's a common situation in molecular biology: genes first identified in a very specific context get named for that, only to turn out to have generalized functions far downstream of that particular role. The (mis)naming of a gene for what it is first observed to "do", or for where it does it, is partly just pragmatic. But it can also be confusing, and surely reflects the prevailing view attested as recently as the late 1990s by cancer specialist Robert Weinberg that "each human gene [is] assigned the role of organizing a distinct body trait." Given that false assumption, it is no wonder the genetics of cancer seemed so perplexing for so long.

have another important mechanism that stops them becoming cancerous. If their genetic systems go awry, they are likely to undergo a kind of self-inflicted death: suicide, if you will. Death is in fact the fate of any somatic cell lineage: cells will automatically die (a process called apoptosis) after they have divided a certain number of times, called the Hayflick limit (see page 108), which for humans is around 50. This makes sense, because each time a cell divides there will inevitably be some mistakes in replicating its DNA: no enzyme machinery is good enough to ensure perfect fidelity in copying two sets of three billion base pairs. So cells that replicate indefinitely will acquire ever more faulty and aberrant genomes. Much better, then, to set a limit and let those degenerating cells die off, while replenishing the tissue by production of fresh lineages.

Apoptosis also commonly happens for cells that lose their proper role in the growth and maintenance of the body. It takes place, for example, in cells grown in a petri dish that become separated from the main mass – a reason to suggest that cells are inherently "social" entities, acquiring their raison d'être only in relation to others and depending on signals from their neighbours to sustain them. "Apparently the only thing our cells do on their own," says biologist Martin Raff, "is kill themselves, and the only reason they normally remain alive is that other cells are constantly stimulating them to live."

Cell death is also involved in the sculpting of the body from growing tissue: it is what removes flesh from between the separate digits of the growing hands and feet, which begin as paddle-like masses in the embryo. If this fails to happen, some fingers or toes may remain connected by an intervening web, as they are (for good adaptive reason) on the feet of ducks.

But cancer cells can evade apoptosis and go right on dividing, so that a tumour grows and grows. They do so by undermining the built-in "division counter" that cells possess. This counter takes an ingenious form. At the tips of our chromosomes are stretches of DNA called telomeres, which don't encode proteins but are merely end caps. Each time the chromosomes are replicated, the telomeres are incompletely copied and get progressively shorter. When the telomeres are completely worn away, things go wrong with the unprotected tips of the chromosomes – they may fuse with one another, for instance – and the resulting chaos prevents the cell from growing and primes it for death.

Cancer cells avoid this by manufacturing an enzyme called telomerase, which repairs telomeres. Cells in the early embryo need to conduct such repair too, which is why our cells *have* telomerase genes in the first place – but usually they are mostly silenced thereafter.* In cancer cells, these genes are reawakened.

Thus there are many requirements that cells must satisfy before they develop into a malignant tumour – it's not simply the case that any old bit of DNA damage will set that fateful process in train. The brakes of the cell cycle must be removed to enable unchecked proliferation. The cell must evade tumour suppressors, and they must fire up their telomerase genes to avoid apoptosis, and stay hidden from the immune system (on which, more later). And they must actively induce the body to grow new blood vessels so as to keep cells deep within a tumour supplied with nutrients. They must eventually be able to spread throughout the body in the lethal process called metastasis.

All this can easily sound like a systematic and purposeful plan. It's no wonder, then, that cancer cells are so often cast as rogues, renegades and killers. Some researchers express this in remarkably

---

* Some immune cells are unique among normal somatic cells in being able to reactivate telomerase genes too, because the immune system depends on being able to renew its cells.

teleological terms, talking of pre-cancerous tumour cells employing all manner of dastardly "tricks" in a selfish effort to ensure their unchecked proliferation, as if these cells are intent on "outwitting the system". Meanwhile, the natural defences of the cell are depicted as labouring valiantly to prevent this disaster: the p53 gene, for example, is in cancer expert Robert Weinberg's words "the arbiter between life and death, an ever-watchful guardian that monitors the cell's well-being and sounds the death alarm if the machinery of the cell is damaged." Quite a responsibility.

There's a pedagogical value in such language that chimes with our own sense of cancer as a terrifying and deadly enemy. But it's worth keeping the metaphor always in view as no more or less than that. It may be that on other occasions, for other purposes, it is better to accept cancerous growth as something innate to cells rather than as a pathological deviancy. It surely results, after all, from the same evolutionary "imperative" that gave rise to us in the first place. It reminds us that there is a contingency in our emergence and continued existence as a cell community. As individuals and as generic human forms, we are just one of the possible outcomes from the potential that exists within our formative cells as they multiply and communicate.

* * *

I said that one of the things cancer cells must do is avoid the immune system. Within this narrative, our immune cells are the body's police, patrolling the body on the lookout not just for invaders like viruses and bacteria but "antisocial" cells of our own. If cancer cells are seen as the villains, immune cells are the heroes.

But according to today's understanding, the immune system does much more than this. Sure, it includes cells called lymphocytes that will identify, catch and digest invaders. But it also plays an important role in many diseases that are more to do with malfunctions of the tissues than with pathogens: brain and heart diseases, obesity, arthritis, diabetes. In part, that's because the immune system controls what is almost the default response to disease or malfunction in the body: inflammation and repair. According to immunologist Lydia Lynch, "we can't ignore the immune system in any disease."

Few areas of biology have advanced more in the past several decades than immunology. At the same time, few are more forbidding to outsiders, more bristling with impenetrable acronyms and hedged with exceptions and complications. The traditional picture was at least fairly straightforward in broad outline: the immune system was considered to fend off pathogens such as bacteria and viruses by generating a huge variety of specialized white blood cells called B and T cells, with "receptor" proteins on their surfaces among which one of them might have the right shape to latch on to the intruder. When that happens (the story went), the machinery of the immune system kicks into gear to eliminate the foreign particle, for example by recruiting cytotoxic or so-called killer T cells to kill and ingest the infected cells. Meanwhile, T cells are "trained" to recognize the body's own cells from the molecules called HLA proteins (never mind the abbreviation – it's another misnomer) on their surfaces. No one's HLA proteins exactly match anyone else's, but they fall into general classes that determine the compatibility of tissues for transplants.

This is all more or less right, but is far from the whole story. For one thing, the generation of B and T cells with diverse receptors constitutes only one part of the immune response, called the *adaptive* immune system. In the late 1980s, immunologist Charles Janeway proposed that there is also an *innate* immune system that deploys immune cells with standardized receptors for common pathogens. Other researchers verified Janeway's bold idea, for which a Nobel prize was awarded in 2011 – sadly, too late for Janeway, who died in 2003. The innate immune system has its own suite of "killer cells", while being able also to awaken the adaptive immune system. It is the first line of defence, and is an evolutionary older strategy, still dominant in plants, fungi and other primitive multi-celled organisms. It has the advantage of being much faster to respond, albeit at the cost of lower flexibility for recognizing new types of pathogen. The adaptive immune system, meanwhile, main-tains a memory of past encounters and thereby confers the immunity on which vaccination depends.

A major aspect of the immune response is tissue inflammation, a signal of the body's defences going into action. However, it's all too common for the immune cells to respond to false alarms, leading for example to allergic responses, with all the accompanying discomfort of mucus generation and skin irritation as the body tries to destroy or expel foreign entities that actually pose no threat. Immunity can also fail in the essential task of discriminating "friend" from "foe", so that it attacks the body's own cells in disor-ders of "auto-immunity" that include rheumatoid arthritis, type 1 diabetes and muscular dystrophy.

Such dysfunctions are normally avoided by a suite of immune mechanisms. Some types of T cell, for example, regulate the immune response, shutting it down once it has done its job. A part of this regulation involves a protein called CTLA-4, production of which is switched on after the T cells have been activated by detection of some foreign agent. CTLA-4 issues the instruction "ease off now", acting as a brake or so-called "immune checkpoint".

One of the most exciting and promising strategies for treating cancer in recent years enlists the immune system to identify and selectively destroy tumour cells. Cancer immunotherapy manipu-lates these checkpoint molecules, blocking their function so as to take the brakes off the immune response and unleash its full force on cancer cells. It uses protein drugs designed to latch on to specific target molecules on T cells and turn off their immune checkpoints.

Directing the immune system at cancer is a very old idea, but the checkpoint approach shows signs of really working, in some cases producing long-lived remission. It was pioneered in the 1990s by biologists James Allison of the University of California at Berkeley, who focused on manipulating CTLA-4, and Tasuku Honjo at Kyoto University, who worked on another checkpoint gene called PD-1. (Different brakes might be more effective for different types of cancer.) Allison and Honjo shared the 2018 Nobel prize in medicine for their work.

The challenge in using immune cells to fight cancer is that cancer cells are not, of course, your regular kind of invader. They are our own cells, which have switched to a state that endangers the organism as a whole. If the immune system has painstakingly acquired an ability to ignore the body's native cells, how can it attack cancer cells? But in fact the immune system does possess some ability to distinguish cancerous from healthy cells. The changes that make a cell tumour-forming may already be enough to set off alarm bells for the patrolling T cells. At the same time, cancer cells acquire some ability to evade surveillance. It's a finely poised game of hide-and-seek, in which efforts to intervene with drugs or agents that make cancer cells more visible to the immune system run the risk of triggering other, harmful autoimmune responses too. One approach under development involves genetically modifying a patient's T cells so that they will attack tumour cells specifically, leaving others alone. Another is to tweak the microbes in a patient's gut (see below) to boost the power of the treatment: certain gut bacteria make the immune system more responsive, but not everyone's gut hosts them.

A checkpoint suppressor drug called Ipilimumab is now licensed by the United States' Food and Drug Administration (FDA) for use against skin cancers. A course of treatment costs over $100,000 but, especially when used in conjunction with other immune-boosting drugs, the results have been encouraging. Cancer immunotherapy has in the past few years begun to make a real impact on treatments in the clinic, with some patients given these treatments making remarkable recoveries from conditions that would previously have been fatal. One woman recovered from a lung tumour the size of a grapefruit, and a six-year-old child was rescued from near-terminal leukaemia. Cancer immunotherapies seem to have particular promise for treating skin and blood cancers (melanomas and leukaemia). Researchers at the Fred Hutchinson Cancer Research Center in Washington state have reported initial clinical trials in which more than half of the patients treated went into complete remission, and for one particular kind of leukaemia, fully 94 per cent of the patients – most of whom would have been diagnosed as terminally ill – saw their symptoms vanish. According to Lydia Lynch, thanks to cancer immunotherapy, "we are now using the taboo word [in cancer research]: cure."

* * *

Both cancer research and immunology show how fragile our sense of somatic integrity is. We might feel that we are unified beings making our way through an environment beset with external threats, but at the cellular level our health and existence depends on a constant and immensely complicated transaction between a diverse panoply of entities whose roles, for better or worse, may depend on their context as much as their content. And if it is not already disorienting enough to appreciate what our own cells are up to, we must factor into the equation the fact that around half* of the cells in our body are not "ours" anyway.

---

* This proportion used to be set much higher, but it now seems that was an overestimate.

They are the cells of organisms that exist symbiotically with us, living all over and within our flesh in an arrangement that is mostly of mutual benefit. The most familiar component of this so-called *microbiome* is the bacteria within our guts, which (among other things) assist digestion: these are the "good bacteria" that we are encouraged to supplement with probiotics and yoghurt. But there are also micro-organisms on our skin, in our mouths, and in many parts of our bodies – not just bacteria but also fungi (such as yeasts) and the other type of prokaryotic single-celled organisms called archaea. In all, these microbial symbionts account for several pounds of our body mass.

Our own cells have adapted to live with these companions: not just to rub along, but to help one another. The gut bacteria are sufficiently important for wellbeing that a mother's milk contains sugars produced especially to nourish them, which the infant itself can't digest. The gut bacteria, in return, assist not just in digesting food but in building and repairing the body, for example by replenishing the gut lining and participating in the process by which the body stores fat. In some organisms, even fundamental biochemical processes may be outsourced to symbiotic microbes. Termites and some cockroaches eat wood but can't digest it; that is done using the enzymes produced by their gut bacteria. The Planococcus mealybug makes some of its vital amino acids using enzymes from two types of symbiotic bacteria.

It seems likely that the very growth of the embryo is partly orchestrated by the microbiome: these symbionts have been shown to activate certain genes needed during development in a variety of organisms including the zebrafish, fruit fly and mouse. Proper development of the mouse immune and digestive systems, for example, depends on chemical signals delivered by bacteria. Even fitness, mating and reproduction may be under the control of symbionts: bacteria in fruit flies make pheromones that influence the host's mating preferences.

There are likely to be interactions of this kind involving our own microbiome too, further undermining the idea that our genome is a self-contained "instruction manual" for constructing a properly functioning human. When such higher organisms are grown in laboratory conditions that suppress their microbiome entirely, the resulting creatures typically have precarious health.

The microbiome's role in our health is profound and perhaps even a little disturbing. The balance of our internal ecosystem is commonly altered by disease, and while it can be hard to distinguish cause and effect, there's some reason to believe that tinkering with the microbiome could lead to cures or at least to alleviation of symptoms. The human gut is connected to the brain by a long nerve called the vagus nerve: a conduit that might give the micro-biome some influence over our mental state. Some research suggests that probiotics – particular types of beneficial bacteria – could be used to treat conditions such as stress and depression, while the gut microbe *Bacteroides fragilis* has been implicated in some of the symptoms of autism. Such claims remain contentious, but it is by no means absurd to think that our microbiome has an influence well beyond a good digestion.

Our body's microflora is highly personalized: no one else's will be quite like it. What's more, this ecosystem, like that of a jungle, varies over the territory it occupies – even the microbes on our left hand will be different to those on our right. And the function a microbe has may depend on where it is: a bacterium might be beneficial in the gut but pathogenic in the bloodstream. As with most stories in biology, context is everything.

Evidently, this collaborative community of non-human cells has to be tolerated by the immune system. In fact, the immune system might be engaged as much in managing the microbiome as it is in warding off hostile invaders. What's more, our body's microbes appear in turn to exert some influence over the mechanisms of immunity – commandeering it, you might say, to protect themselves against non-indigenous microbes.

Given this intimate and essential interaction between our human cells and those of the organisms we host, we need to respect and nurture the microbiome. We know all too well how rotten we can feel if it is knocked out of kilter after a course of antibiotics (although that's by no means the only reason for the side-effects of those drugs). All the more reason then to be wary of a culture that sees fit to coat every surface with bactericidal agents – the vast majority of bacteria are, after all, harmless to humans. At the same time, the many contingencies of the microbiome should warn us once again against making any simplistic taxonomy of "good" and "bad" cells. As science writer Ed Yong has said:

> There is no such thing as a "good microbe" or a "bad microbe".
> These terms belong in children's stories. They are ill-suited for describing the messy, fractious, contextual relationships of the natural world.

The close connection between a host and its microbial symbionts, in which both contribute to the survival of what we might call the "joint organism" – which some dub the holobiont – complicates evolutionary theory. If a microbiome actively participates in the holobiont's survival prospects, what then is the entity on which natural selection acts? But a microbiome is not inherited intact: yours will share some features with your mother's but is not identical. Some of the microbes hosted in your birth mother will have been transmitted to you during passage down the birth canal, and also subsequently during physical contact and breast-feeding. But some will have been acquired from elsewhere, including other individuals around you during infancy. So it's not clear how to incorporate the microbiome into evolutionary thinking: should it be considered a genetic pool entirely separate from that of the host, or at least partly connected to it?* Evolutionary biologists are divided over the issue. Perhaps, suggests one of their veterans W. Ford Doolittle, the coherent evolving entity is neither host nor microbial symbiont but the *process* they enact together: the pattern of metabolism, say. Doolittle likens this to the way songs are perpetuated (while evolving!) by people singing them: it's the song that survives and evolves along the way.

Some researchers even go so far as to suggest that we should think of organisms possessing a "hologenome" constituted by the genes of the host and its symbionts. Others dismiss or even deride that idea – although, as we'll see later, it is far from obvious how biology can be parcelled up into individual organisms. At any rate, there's nothing in this blurring of genomic selfhood that is incompatible with the standard Neodarwinian view of evolution. The question is whether or not it constitutes a useful way to tell the story – or rather, perhaps, of what kind of story you're trying to tell.

---

* It seems likely that some genes from an organism's microbiome can find their way into the genome of the host.

CHAPTER 4

TWISTS OF FATE

HOW TO REPROGRAMME A CELL

It was just a little pink lump, that scrap of flesh scooped from my upper arm with a surgical instrument. I was too squeamish to watch it being carved out, but in the test-tube it reminded me of the parings I'd accidentally remove from my fingers as a teenager, when my hand holding the razor-edged craft knife slipped as I assembled my model soldiers and tanks. There's something horribly fascinating in seeing our flesh reduced to a kind of plastic material, yielding under the blade before the blood wells into the incision.

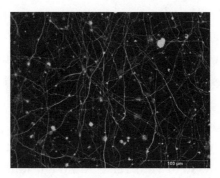

My neurons growing in a layer in vitro.

This piece of me was going to become not just a mini-brain but the source of other neurons too, cultured in flat layers as they spread tendrils that fluorescent staining agents turned into designs of considerable beauty.

Not much more than a decade ago, many biologists would have considered these things impossible. To be the fabric of my arm was considered the final and irreversible fate of that particular cell lineage. All they could do subsequently was renew themselves, each time leaving the tissue a little the worse for wear, its elasticity compromised in a cartography of wrinkles, until my entire body gives up the ghost. But no: here they were, given a new lease of life and a new identity. It seems it is never too late for cells to start afresh.

This revolution in our view of what cells can do is at the heart of this book. More than the mere *culturing* of cells ever did, the *reprogramming* of cells cuts the tethers to our notion of biological time. Instead of a single life story, it reveals that the embryo from which we grew has no limit to its narratives. Our flesh, you might say, has been liberated. We're working out what to do with that freedom.

* * *

For the first cells, anything is possible.

We saw earlier that the single fertilized cell, the zygote, from which we all begin is totipotent: able to develop into every type of tissue needed by the growing embryo. Little by little, the dividing cells lose this versatility along the road to making us.

Some cells specialize to become the placenta that nourishes the embryo, while those that make up the embryo proper stay pluripotent, able to form all tissue types in the mature human body. These are the embryonic stem (ES) cells. Although long known to exist, mammalian embryonic stem cells weren't identified and isolated until the 1970s and '80s. Biologist Martin Evans, working at Cambridge

University, succeeded in culturing them from the blastocyst embryos of mice in 1981,* for which he was awarded a part of the 2007 Nobel prize in medicine. The challenge in culturing stem cells is to prevent them from differentiating into a particular tissue type, and it's a little like trying to balance a needle on its tip. It was first achieved by growing the cells on a layer of others that have already acquired their fate, such as fibroblasts, which "feed" the stem cells with a protein needed to sustain their pluripotency. This protein has since been identified and can be added directly to the culture medium.

It wasn't until 1998 that James Thomson of the University of Wisconsin-Madison, and independently John Gearhart of Johns Hopkins University in Baltimore, succeeded in isolating and culturing human ES cells. Thomson took them from IVF embryos grown to the blastocyst stage, while Gearhart grew them from early-stage germ cells extracted from aborted fetuses.

By the time an embryo has grown into a pre-natal baby, the ES cells have vanished, although some stem cells with a reduced repertoire remain in the blood of the umbilical cord. A few cells in our bodies retain a degree of "stemness" too, being capable of regenerating several of the cell types needed for specific tissues. Most tissues contain these "mature" (so-called somatic or adult) stem cells; they step up to the plate to replenish flesh during wound healing, for example. Adult stem cells in bone marrow, called hematopoietic stem cells, can generate the different types of blood cell in the body, and have been used for many years in transplants to replenish blood-forming cells after they have been damaged by chemotherapy in leukaemia patients. Some of these become lymphoid progenitor cells, which then specialize into the lymph cells of the immune system, such as different types of

---

* Much of the groundwork for this achievement was done by biologist Gail Martin, who figured out how to isolate and sustain a kind of stem cell from tumours of germ cells.

lymphocyte. Other hematopoietic cells become myeloid progenitor cells, which can specialize further into red blood cells, basophil cells involved in inflammation, clotting and allergic responses, and neutrophil cells, the main white-blood cell type involved in immunity. Adult stem cells in the brain, meanwhile, can produce the many distinct types of cell found therein: both neurons and non-neuronal cells such as so-called glial cells.* These tissue-specific stem cells allow certain tissues to regenerate constantly: skin cells typically last just a month or so, blood cells about four months. Some adult stem cells, such as those that make muscle, are activated only in response to particular environmental signals like injury or trauma. Quite how such semi-specialized stem cells resist the pressure to fully differentiate, as their neighbours have done, is still not entirely clear.

So adult stem cells give us some capacity to regenerate tissues. But it is very limited. We can't regrow a lost limb, as the salamander can. Here it is, going happily about its amphibious business lakeside, when along comes a heron and snap! – it has lost a leg. No problem. It scuttles away to safety and its cells kick into action, covering the wound with a mass of proliferating tissue that, before long, has grown into a replacement for the missing limb.

How enviable that is. What if we could grow a new kidney to replace one damaged by disease or injury? Or new nerve cells in the spinal column to recover from a paralysing fracture of the backbone? Might we even imagine regrowing all the various tissue types of a finger, a hand, a leg?

Regeneration of lost or damaged tissues has been on the agenda ever since scientists began to understand the properties of stem

---

* The regeneration of neurons in humans was discovered only recently by researchers at the Karolinska Institute in Sweden. In 2013, they used radiocarbon dating in the human body to show that some neurons in the brain's "memory centre", the hippocampus, are relatively young. About 700 new neurons are thought to arise in the hippocampus each day.

cells in the 1970s. Already human embryonic cells, taken from embryos discarded in IVF fertility treatments or saved from umbilical blood, have been cultured into tissues such as "artificial" skin for grafting to burns victims. Some attempts to use stem cells in medicine deliver them surgically to where they are needed, in the hope that the tissues will guide them to the right fate. Other stem-cell therapies still being trialled inject these cells directly into the bloodstream, trusting them to find their way to the intended target. Stem cells have ways of doing so, if they are already on the road to a particular fate such as muscle or heart tissue.

But there are many obstacles, which is why stem-cell therapies have largely failed so far to deliver on their early promise. For one thing, it can be challenging to control the fate of stem cells so that they develop into the desired tissue. Indeed, one of the big concerns about using ES cells, which are fully pluripotent, in medicine and surgery is that they may acquire the wrong fates, giving rise to tumorous growths. And stem cells taken from embryos or donors and transplanted to patients are liable to be rejected by the immune system, just as in any regular organ transplant. What's more, many human organs contain several different types of cell juxtaposed in complicated patterns and structures that we don't yet know how to replicate. (We'll see in the next chapter that we are learning.)

There are ethical problems too. In some countries, the use of ES cells is highly controversial, because of objections to research involving (and leading to destruction of) human embryos. Such concerns led US president George W. Bush in 2001 to ban the use of newly created human ES cells in federally funded research, greatly hindering scientific work on stem-cell medicine and therapies.* There are similar restrictions in Japan.

---

* There were actually some restrictions already in place since 1996 on the use of US federal funds for research involving human embryos. This meant that, for

It was because of obstacles like this that some researchers wondered if there was another way to get stem cells that didn't involve human embryos or fetuses. How might that be done, though, if pluripotent stem cells exist so fleetingly before becoming committed irrevocably to some specialized fate?

*   *   *

Until just a decade or so ago, most biologists assumed that a cell's fate was a one-way journey. Once a stem cell had become committed to a particular lineage, it was believed that there was no going back. We saw earlier how, in a differentiated cell, some genes are switched off while others might become more active, enabling the cell to develop and function in the way needed for that specific tissue. Many researchers thought that, as genes became redundant in differentiation, they were permanently lost – not necessarily excised from the genome but deactivated for good.

In 1928, Hans Spemann began to explore how the activity of chromosomes becomes increasingly constrained down a cell lineage. Using fine tools under the microscope, he extracted the chromosomes from a fertilized salamander egg, and then moved the nucleus (containing other chromosomes) from another embryonic salamander cell into the emptied egg. He found that the egg would still grow into a complete embryo, guided by the genetic programme of the transplanted chromosomes. This "nuclear transfer" makes the new cell a clone of the one from which the nucleus is taken: Spemann's experiment was the first example of intentional cloning of a complex animal.*

---

his breakthrough research on the culturing of human ES cells, James Thomson had to rely on private funds from Californian biotechnology company the Geron Corporation, and had to perform most of the work himself in an off-campus lab at Madison.

* You could also argue that his experiments on separating cells in a salamander

How far down the developmental road does that potential of chromosomes to reprogramme a cell persist? In 1938, Spemann speculated about doing the same experiment using a fully differentiated adult cell – a somatic cell, taken from some body tissue – as the nuclear donor. He wasn't able to carry out the experiment but proposed it merely as a "fantastical" idea.

As the history of science repeatedly shows, one person's idea of a fantastical thought experiment may become another's research agenda. In 1952, Robert Briggs and Thomas King of the Lankenau Hospital Research Institute in Philadelphia took a cell from an embryo of the northern leopard frog that had grown to the stage called a blastula (more or less equivalent to a mammalian blastocyst), containing many thousands of cells – by which time the cells were starting to specialize – and transferred its nucleus to an unfertilized egg cell stripped of its own nucleus. The egg grew into a fully formed tadpole, showing that, even once differentiation had begun in earnest, the chromosomes could retain pluripotent status.

There's a feature of these experiments that is often glossed over but which should give us pause for thought. The egg cell that gained the new nucleus and grew into a tadpole *was never fertilized*. All that was needed to start it dividing and becoming an embryo was a little prick from a fine needle: a mechanical prod into action. On the one hand, that's no big deal. An egg cell can't generally go it alone because it is haploid: it only has one set of chromosomes. It needs a second set from the sperm. But the nucleus of an embryonic cell is already diploid, and so transferring that nucleus to a "de-nucleated" egg gives it all it needs. On the other hand, the ease of triggering the genesis of an embryo from an egg that has been given a full set of chromosomes makes fertilization itself – the

---

embryo using a "noose" (see page 57) were also a kind of cloning, producing genetically identical "twins" artificially from a single embryo.

cell's-eye version of sex – seem rather unimportant. As we saw, there's more to fertilization than the arrival of an extra set of chromosomes. But, it seems, not so very much more.

The experiment of Briggs and King showed that embryonic cells can remain pluripotent in principle for some way along the route towards a developed embryo. But there are limits. If the two researchers used the nuclei of embryonic cells at a slightly later stage than the blastula, called a neurula, the egg didn't develop properly. Many biologists concluded that the chromosomes of cells gradually and irrevocably lose their pluripotency the later they arise in the cell lineage. Maybe they even lose some of their DNA?

This remained the common view even after British embryologist John Gurdon had shown that it wasn't always the case. As a graduate student in the department of zoology at Oxford University in the late 1950s, Gurdon carried out the kind of experiment that Spemann had fantasized about two decades earlier: transfer of the chromosome-laden nuclei of somatic cells. He used African claw frogs, which have the convenient trait that they can be made to lay eggs on demand by an injection of a hormone. As the nuclear-donor cell, Gurdon first used differentiated skin-like cells in an embryonic frog's gut. He found that the recipient eggs would still develop into a full-grown frog. Again the resulting organism, bearing the genes of the donor cell, is a clone. Gurdon later carried out the nuclear transfer from a range of embryonic tissue types, such as heart and lung cells.

But still Gurdon wasn't able to achieve the same result using somatic cells from *adult* frogs. Some eggs that received such nuclei would become tadpoles, but they could then develop no further.

Some researchers were sceptical that the differentiated donor cells in Gurdon's experiments were truly and solely responsible for development of the eggs. They wondered if there was a trace of genetic material left behind in the ostensibly emptied egg cell that helped the process along. What's more, although the donor nucleus might be able to direct embryonic development, it sometimes showed a kind of memory of its previous host cell: muscle-cell nuclei, for example, would keep making more muscle-specific proteins than an embryo should. So it wasn't clear what Gurdon's nuclear-transfer experiments said about the capacities of chromosomes in mature somatic cells.

It wasn't until the 1990s that pluripotency of adult chromosomes was clearly established by somatic-cell nuclear transfer. In 1997, researchers at the Roslin Institute in Scotland, led by Ian Wilmut and Keith Campbell, transferred the nucleus from a mammary cell of an adult ewe to a de-nucleated sheep egg, from which they grew Dolly the sheep. The fuss around this first cloning of a large mammal from adult cells – with all the science-fictional narratives it suggested – rather obscured the implications of the result for our picture of cell reprogramming: Dolly showed that mature somatic cells do still have all the genes needed to make a complete organism, and that they can become active again.*

Egg cells evidently possess some inherent ability to reactivate this pluripotency in a somatic-cell nucleus. Once the egg starts to develop, it will make embryonic stem cells with a genetic constitution identical to that of the nuclear donor cell. These could be harvested from the embryo at the blastocyst stage for tissue repair: a technique called therapeutic cloning. South Korean biologist Woo Suk Hwang, working at Seoul National University, claimed in 2006 that he had achieved this with human cells, but his high-profile publications were later found to be based on faked data – for which Hwang was prosecuted and given a suspended prison sentence. In such high-stakes science, misdemeanour has become a constant hazard. No one has yet genuinely done what Hwang had claimed.

But even if human therapeutic cloning were shown to be possible,

---

* There is a small but important exception. A few genes in mammal cells are

it is beset with problems as a source of stem cells for medicine. For one thing, human eggs are not easy to come by. And you still produce a human embryo this way, whereas plenty of people object to that being done purely as a means to an end. Several countries forbid the procedure even in principle.

* * *

Gurdon's experiments suggested, and the cloning of Dolly confirmed, that even mature differentiated cells retain the genetic capacity of stem cells. Normally this capacity is shut down by epigenetic gene regulation, which switches some genes off as differentiation proceeds. It was widely thought that somatic cells need to experience something rather drastic to reawaken those inactive genes. Dragging a nucleus into an egg cell is certainly a pretty traumatic affair. This cell compartment isn't just like a free-floating blob of oil in the cell's vinaigrette but is delicately woven into its fabric: Wilmut and colleagues compared the transfer to a brain transplant. And to find itself inside an embryonic cell can be a rude shock for a somatic nucleus. For one thing, the cells of zygotes divide much faster than those of mature somatic cells, and sometimes this is too much for the replicating chromosomes to endure without damage.

But might a somatic cell's genes be reset without having to pull out its nucleus, stick it in an emptied egg and grow it into an embryo? For most researchers before the turn of the millennium, that seemed about as likely as reversing time – or as likely as transforming you back to your infant self.

Unlikely ideas are often pursued by unlikely candidates. Japanese biologist Shinya Yamanaka was one of those. He took an

---

contained not in the nuclear chromosomes but in the mitochondria, which are where energy-storage molecules are produced. These mitochondrial genes are inherited solely from the mother.

unconventional path to cell biology that made him apt to ask the kind of questions others would have dismissed out of hand.

Yamanaka was a keen rugby player, which understandably gave him an interest in sports injuries. But despite training as a surgeon in the 1980s, he found he wasn't terribly good at it (surgery, not rugby). In 1989, he went to Osaka City University Medical School to take a doctorate in basic medical research, which led him to study heart diseases caused by clogging of arteries, and potential gene therapies for them. By and by, Yamanaka found himself investigating genes that affect the growth of tumours, in the course of which he began to use embryonic stem cells from mice. He discovered that the tumour-suppressing gene he was investigating seemed important for maintaining mouse ES cells in a pluripotent state.

Yamanaka was frustrated by the strong restrictions imposed in Japan on human stem-cell research because of its reliance on human embryos. Might stem cells be produced *without* embryos, he wondered? If the genes responsible for pluripotency in stem cells could be activated in mature somatic cells, might they adopt a stem-cell-like state? Yamanaka decided to give it a try, simply by giving somatic cells fresh copies of the relevant genes.

Thanks to research on gene therapies, there was already a suite of methods for adding new genes to cells. In gene therapy the aim is to overwhelm disease-causing mutant forms of genes in a person's genome by injecting their cells with versions of the properly functioning gene. One of the best ways of getting the new genes into a cell uses viruses as the transporting agent.

As we saw earlier, viruses are parasites operating at the threshold between the living and the inorganic. Rather than replicating themselves as cellular organisms do, they are able to piggyback on the replicating machinery of the host organism's cells. A virus will smuggle its genetic material inside the host cell and weave it into the host's DNA so that, when the cell divides, the virus too gets replicated.

There's nothing malignant in that process beyond what we read into it. Viruses don't of course "intend" to cause problems for their host. They exist because they can: because what they do turns out to be evolutionarily stable. Indeed, some viruses coexist with their host without any adverse consequences. In particular, viruses that attack bacteria, called bacteriophages, can help to ward off bacterial infections. Others might combat one another: the so-called GB-C virus interferes with the activity of the particularly nasty viral pathogen HIV, the virus that causes AIDS in humans. And when viruses do cause health problems, it's often because of the body's response rather than anything intrinsic that the virus does. All that fever and sneezing that follows infection with the influenza virus is caused by the body's defences, such as inflammation, as the immune system swings into action.*

Commandeering viruses for medicine might sound perilous. But for gene therapies, the key attraction of viruses is their ability to get genetic material into cells. With a little genetic engineering, certain viruses can be rendered harmless while retaining their gene-smuggling capabilities. Researchers developing gene therapies use them as "gene vehicles" or *vectors*.

Yamanaka worked out that he could use such viruses to introduce into somatic cells the genes prominently expressed in ES cells. The idea was that the genes might reprogramme the cells into this atavistic state. Once the genome is reset, this should be carried over to daughter cells produced in subsequent cell division.

The problem was that there are *hundreds* of genes that are particularly active in ES cells. Would reprogramming require all of these to be added? Or would just a few be enough? The only way to find out

---

* HIV is somewhat different: the virus does actually attack the immune system, killing off a type of white blood cell that protects against infection and thereby leaving the body exposed to secondary infections.

was by trial and error – but the number of possible permutations of all the respective genes was enormous. "We thought at the time," said Yamanaka, "that the project would take 10, 20, 30 years or even longer to complete." And even then there was no guarantee that this rebooting of cells would work at all. Many people would have simply given up.

It was possible to narrow the list of candidate genes a little, however, because not all of those active in ES cells appear to be equally important for pluripotency. One that is goes by the name of Oct4, and it seems to inhibit cell differentiation: if researchers use gene-manipulation techniques to silence Oct4 in ES cells, they differentiate. It's still not entirely clear how Oct4 plays this role, but it is known to be an example of a *transcription factor*: it encodes a protein that influences the activity of other genes, specifically the rate at which they are transcribed. To perform this function, the Oct4 protein pairs up with another protein from a gene called Sox2. Together they bind to DNA in a way that regulates another transcription factor called Nanog, which is also central to maintaining ES cells in an undifferentiated state. Nanog is a so-called "master gene", and plays a key part in a cell's transition from a totipotent to a pluripotent state.

All this sounds rather abstract, doesn't it? This gene or that gene does something to another gene, and that interaction is mysteriously implicated in a change to the state of a cell.

But that's the point. Remember how inscrutable, in general, is the link between what proteins actually do and their knock-on effects on the phenotype. We can say that Oct4, Sox2 and Nanog are part of a gene-regulatory network that maintains ES cell pluripotency – or perhaps better, that inhibits the cell's default tendency to specialize. If you want to tell a clear, simple story about this process – comparable, say, to the well-defined roles that editors, copy editors and printers perform in making a book – you're likely to be disappointed.

Perhaps it's more helpful to think of the process that determines cell fate as a kind of computation. There are input signals, such as the action of genes like Oct4, and there are outputs, such as the functions that a given cell performs: making insulin in pancreatic cells, say, or producing electrical impulses in cardiac muscle cells. For human embryogenesis, the first input signal that an egg cell receives is the ingress of a sperm, along with prompts supplied by the maternal tissues. As an embryo grows, signals continue to come from surrounding cells: for example, the positioning signals of diffusing proteins that tell a cell where it is and what it should become. These signals tweak and modify the genetic programme of a cell – eliciting, for example, epigenetic modifications that constrain the cell's activity. By injecting transcription factors into a cell, Yamanaka was creating another kind of input signal, biasing the cell's internal computation to generate a specific kind of output.

Explaining the mechanism, the *logic*, that leads from inputs and outputs is still tremendously challenging and might often prove to be more or less impossible in a narrative sense. While the action of just a few genes may be crucial at any given point, that action conditions a complex web of interactions between many genes, and to all intents and purposes the link between inputs and outputs is a black box whose circuits we can either not access at all or glimpse only in parts, hazily. Cell biologists and geneticists might sometimes draw circuit diagrams or cartoons with arrows to try to indicate sequences of events, but there is often a good chance that these express nothing more than correlations between things measured in experiments, with no knowledge of whether they speak to physical processes between genes and proteins that make change happen. And mistaking those two things is apt to lead to wild goose chases, misnomers, bad stories and false confidence.

Be that as it may, Oct4 and Sox2 were evidently good candidates for Yamanaka's "factors" that might induce somatic cells to revert to a stem-cell-like state. He and his co-workers identified 24 such factors. Another of them, called c-Myc, acts as a kind of master-regulator transcription factor that boosts the expression of many other genes involved in the proliferation of cells. It appears to be a hub in the network of gene interactions that governs cell division. That's why genetic mutations that interfere with the action of c-Myc are common causes of cancer (see page 130). In some tumours, c-Myc is working in overdrive, permitting the uncontrolled proliferation of cells.

Working now at Kyoto University, Yamanaka developed a method for introducing his 24 candidate factors into mouse fibroblast cells using viruses. There was an awful lot to test, and he admitted that "to tell the truth, we did not expect that we had the answer among these 24 factors. We thought we had to screen many more."

His first experiments, conducted with his student Kazutoshi Takahashi, went for broke by just throwing in all 24 factors. To the researchers' surprise, it worked: the mouse fibroblasts showed signs of stem-cell-like behaviour. They had been induced into pluripotency.

That was a dauntingly complex cocktail, though. Through a painstaking process of elimination, Yamanaka and Takahashi were able to induce stem-cell behaviour with just four factors whose genes were inserted by viruses:* Oct4, Sox2, c-Myc and a gene called Klf4. They christened these reprogrammed cells *induced pluripotent stem cells* (iPSCs). The two researchers showed that, if iPSCs were attached to embryos at the blastocyst stage, they would be incorporated smoothly into the growing organism. Yamanaka and Takahashi reported their findings in 2006, and the following year they made iPSCs from human somatic cells. James Thomson at

---

* Even if the viruses used here are ostensibly harmless, there is always a danger that they could introduce other changes into the cells' DNA, potentially making them apt to develop into tumour cells. That's not a problem in the petri dish, but it is if you're planning to use the cells in medicine. So researchers are looking for other ways to get the right factors into somatic cells. One is to add the respective proteins directly, not the genes that encode them. But the efficiency of reprogramming is typically lower in that case: fewer cells are switched to a stem-cell state.

Wisconsin achieved the same thing independently but reported the results somewhat later.

The discovery was hailed as the advent of "ethical stem cells", which do not require any human embryo to be "sacrificed". The reasons for that designation are clear enough, but still it is contentious. First, we have to wonder if iPSCs are truly like embryonic stem cells, or whether they contain any epigenetic memory of their previous differentiated state. That is still debated, but several studies have shown that the epigenetic tags introduced onto the genome in differentiated cells are not fully erased in iPSCs. The patterns of this "epigenetic memory" can differ even for fibroblast cells taken from different parts of the body on a single individual, and may depend on the age of the donor. For medical applications, the question is if and how much these memories matter. If such cells are going to be used to grow new tissues and organs, an epigenetic memory might influence how well they do their new job – you might not, for example, want neurons made from iPSCs behaving a little like skin cells. If we had any plans to use iPSCs for reproductive technologies, for example to make human embryos for implantation and gestation, there's a possibility that these embryos might be in some sense prematurely aged.

What's more, to regard iPSCs as the "ethical option" is already to accept the idea that there is something unethical about using human ES cells for research. That, certainly, was the position taken in 2001 by the committee of bioethicists convened by George W Bush to provide recommendations on embryo research. It was chaired by the arch-conservative Leon Kass, who advocates what he called the "wisdom of repugnance" as a guiding principle: a fancy way of saying that if something makes me feel "yuck!" but I can't articulate why, I should listen to my instincts and ban it. It was no surprise, then, that the panel recommended the banning of federal funding for any research that involved lines of human

ES cells not already in existence (there wasn't exactly much point in prohibiting ones that were). The decision effectively stifled stem-cell research in the United States (which, however, is by no means the only country to adopt such a position).

There are good arguments to challenge such a restrictive and dogmatic position about the moral status of embryos, but that debate is for elsewhere. Suffice it to say, there are no definitive facts that can settle the matter objectively. In purely practical terms, iPSCs might offer a way to study and use stem cells in the face of prohibitive legislation; but there is no guarantee that they are entirely equivalent to embryonic stem cells, and much likelihood that they are not.

John Gurdon and Shinya Yamanaka shared the 2012 Nobel prize in medicine or physiology, and their achievements were hailed as heralding potentially revolutionary advances in clinical medicine and medical research. But the real implications go still deeper. The Nobel citation asserted that their work created "a paradigm shift in our understanding of cellular differentiation and of the plasticity of the differentiated state". To put it less euphemistically, you could say that it transformed the notion of our "being", making it no longer enslaved to the passage of time.

Our whole concept of the emergence of self has always been predicated on the assumption that this process is a one-way street: we are conceived, we grow, mature and age. The elaboration of a single cell into the colony we call a human body seems conceptually unproblematic so long as we can regard it all along as a coherent, boundaried organism with a unique history. But if you can take a tiny part of that mass of cells, rewind the tape and *make a fresh start*, then this unity starts to dissolve.

* * *

Our understanding of a cell's lineage had been conditioned by Conrad Hal Waddington's powerfully intuitive concept of a landscape of states, with its hills and valleys. But we can now see that it is a simplification not just in terms of the supposed one-way direction of travel. For there is not exactly a "landscape" at all.

It's often the case in science that improvements and advances in our ability to study nature raise more questions than they answer, showing us only that our comfortable notions are but crude approximations. That's how it is with cells. It has now become possible to characterize the complete genetic, genomic and transcriptional status of individual cells: how many proteins or RNA molecules of different types they are producing, for example, and the exact epigenetic state of the genomes. Such work has made it clear that in a given tissue, at a given moment, there is typically a wide range of different states of cell activity, and that they can't simply be described by balls rolling down Waddington's smooth and well-defined valleys and ridges.

Systems biologist Allon Klein of Harvard University compares it to the Dutch artist M. C. Escher's famous drawing of a topsy-turvy stairwell that has no unambiguous up or down. "Gene expression space can in some places be a bit like this," he says. "There simply isn't a natural concept of a landscape here that fits. Directions seems to change unexpectedly, and a ball could roll indefinitely around the stairwell." And if all we have are snapshots of cell states, it's almost impossible to be sure how the cell got there. "For any snapshot, there are multiple different dynamic processes that could give rise to the same picture," says Klein.

You could see it as analogous to the way genomics itself is revealing the rich diversity of human populations. It is convenient for some purposes to classify people into national groups (French, Indian, Japanese . . .), and also into what we conventionally call racial groups: Han Chinese, sub-Saharan African, white European, say. But these labels mean next to nothing at the genomic level, where we see tremendous genetic diversity within populations that we might regard for demographic purposes as homogeneous. Someone who is a white Spaniard, say, may turn out to have a very different genetic heritage and genomic profile to another person who ticks the same box in a population census. And we can sub-divide genomic relatedness as finely as we like: there's nothing fundamental about our conventional distinctions of race, nation-ality, family. All the same, these categories can be useful if viewed with appropriate caution.

It's similar for cells. There may be more than one route, say, from an embryonic stem cell to a red blood cell, and cells in the kidneys that we might happily group together as hepatocytes could have rather different profiles of gene activity. The "absolute" nature often ascribed to cell types has been encouraged by the bluntness of the tools traditionally used to look at them. In the images of my mini-brain, for example, or of a developing embryo, the staining chemicals used to label cell types are dyes that become attached to particular proteins. If a cell is making that protein in large quantities, it absorbs the dye. This amounts to making the production of a single protein the determinant of cell type. It's not a bad way of classifying cells, since often the switching on of a particular gene (and the production of its corresponding protein) is a good indicator of the cell's trajectory on the map of possible cell fates. Cells expressing a lot of Oct4 and Sox2 tend to be stem cells, say. Still, that is just one data point in a vast, multi-dimensional space. We have 23,000 genes and around 60,000 proteins, and to fully describe a cell's state we need to find its place in this many-thousand-dimensional space.

Of course, we can't visualize what that means. Nonetheless, scien-tists have now developed the experimental and computational tools to begin that kind of mapping, and they have found that both the state and the lineage of a cell may be more complicated than these all-or-nothing maps of cells glowing blue or red under the micro-scope would suggest.

For example, Klein and his collaborators have mapped out the developmental landscape of cells in zebrafish embryos as they grow from a pluripotent state called a blastomere in the early stage of embryogenesis towards a variety of specialized cell types. Many cells are in somewhat ambiguous states when their entire gene expression profile is taken into account. In place of Waddington's landscape, there are rather broad basins, sometimes conjoined by a complex terrain without a clear ridge separating them, and which may part and rejoin in loops as well as branches. The paths that cell lineages may take are diverse and not always obvious.

That's not to say that there aren't identifiable cell-types at all, mind you. "The mature cells are extremely well defined," says Klein. "And if you look at the genes that give the cell types their final functions, they really are specific to one cell type or another." But he says that cells that haven't yet committed to a fate "form a continuum" – a kind of broad plain on which it may be far from obvious which pathway a given cell is on.

Perhaps we might more usefully replace Waddington's image of valleys with that of a rather flat landscape with cities and towns at its periphery and a dense network of routes connecting them. This "map of cell states" is not chaotic – it has structure and logic. But that structure is complex, multi-dimensional and dynamic, and is only crudely captured by simple, static pictures.

And that is modern biology for you: the art lies in finding useful approximations for thinking about systems shaped by evolution into massively complex and ever-changing structures in which regular and predictable behaviour is emergent and often subject to exceptions and unexpected diversions. That's why we need to choose our metaphors with great care, and to resist being ensnared and bewitched by them.

At any rate, this cell-by-cell mapping of embryos is now offering a dizzyingly detailed view of how the activation of genes gives rise to the formation of bodies. In 2019 a team of scientists at the

University of Washington in Seattle reported the "landscape" of cell states for the mouse embryo as its organs start to appear during days 9 to 13 of gestation: an atlas of cell types and their signature genes for no fewer than 2 million individual cells in the embryo. The sheer volume of data and the technological prowess needed to collect it are astonishing enough. What is perhaps most telling of all, though, is the mere recognition that to fully understand development we will need to probe at this fine-grained level of detail, perhaps at the limits of our ability to visualize the information and to make any sense of it. We are a long way now beyond the broad-brush "organizing fields" of early embryology – but equally, still a long way away from being able to claim a deep understanding of how mice, let alone humans, are naturally grown.

* * *

It was Yamanaka's cell-reprogramming method that Selina and Chris used to transform my fibroblast (skin) cells to iPSCs and thence to neurons. The technique is what makes it possible to grow organoids of many kinds from the somatic cells of adults. Progressing from a stem-cell stage, they recapitulate in a more or less approximate way *in vitro* the growth of the corresponding organ in the developing embryo that became *us*.

When cell biologist Madeline Lancaster produced brain organoids as a graduate student of Jürgen Knoblich at the Institute of Molecular Biotechnology in Vienna in 2013, it was by accident. At first, she didn't know what these white blobs were that had appeared in her cultures of stem cells. Gradually, she recognized the truth: that the cells were trying to make brains.

To grow a mini-brain demands considerable skill in cell culturing, but the cells themselves don't need any guidance. They just get on with it. "It doesn't require any super-sophisticated bioengineering," Knoblich says. "We just let the cells do what they want to do."

"Simply by providing the right conditions to support and nurture the cells in a three-dimensional configuration, the [mini-]brain is able to self-organize and build itself," says Lancaster. "At first it was totally surprising that these cells could make a structure rather like a brain all by themselves," she adds – but in retrospect it makes complete sense. That kind of self-organization is, after all, "just what an embryo does".

All manner of organoids can be grown from both embryonic stem cells and iPSCs, if they are nudged from their pluripotent state in the desired developmental direction. If the cells acquire a kidney-cell fate, they will be predisposed to grow into kidney-like structures; pancreatic cells will form a kind of miniature pancreas, stomach cells become stomach organoids. Sure, these structures are not real kidneys and so on, nor even reduced-scale versions of them. But neither are they an undisciplined mass of kidney cells. The organoids take on some of the real features of the genuine organs in question. The cells may differentiate into some of the specialized types found in the respective organs, and the tissue that they constitute may become shaped into an approximation of the real article via the processes of cell migration and sorting that happen in a developing embryo. That, for example, is how stem cells taken from the intestines of mice spontaneously organize into hollow compartments lined with the protrusions called villi that are adept at absorbing nutrients. It's what enables human neurons growing into mini-brains to separate into the distinct layers seen in the cerebral cortex and to form structures reminiscent of other parts of the brain and associated organs, including primitive light-sensitive retinas.

All the same, organoids need a little help. The game-changer for making mini-brains, says Lancaster, was a protein-based gel called Matrigel that acts as the soft, supporting and in some sense *encouraging* medium in which the neurons grow. Such work was pioneered by Japanese biologist Yoshiki Sasai, who reported in 2008 that he and his co-workers had grown neuronal structures resembling brain regions – forerunners of mini-brains – from mouse embryonic stem cells in a carefully formulated gel medium. The following year, other researchers grew "gut organoids" from adult human intestinal stem cells. From 2011, researchers began to use human induced pluripotent stem cells as the source instead.

Lancaster's brain organoids made their debut in 2013. But Sasai, whose work was so vital to the field of organoid growth, committed suicide the following year after a scandal involving scientific misconduct in the team in which he collaborated. Those researchers had reported an astonishing new way to induce stem-cell behaviour in mature mouse cells: not by injecting them with transcription factors but just by subjecting them to stress, for example exposure to acid. The experimental reports, published at the start of 2014 but now discredited, were shown to have included manipulated data. Sasai himself was cleared of any tampering with the results, but he was deemed to have failed in duties of oversight that allowed the more junior lead author of the papers to get away with fraud. He was harshly criticized in the Japanese media and accused of chasing aggressively after large grants to fund the work. The pressure was too great, and in August of that year he was found dead, having left several suicide notes.

* * *

Organoids grown *in vitro* might one day constitute good enough approximations of the real thing to be used in transplantation, replacing damaged or malfunctioning tissue. They might offer personalized, compatible structures for treating ailments like kidney failure, heart disease or even neurodegeneration. Perhaps we will grow new body parts as easily as we (some of us) grow hair and fingernails.

But that vision is still some way off – and in the next chapter I look at other ways in which artificial tissues and organs are being grown. Left to their own devices in a petri dish, iPSCs and ES cells do a rather indifferent job of making organs. There seemed something almost poignant in the images of other mini-brains that Selina and Chris showed me. Some of these bristled with hopeful little nodules, sprouting in search of the signal anticipated from surrounding tissues which would indicate where to lay down the nerve bundles of an incipient spinal column. No such guidance was to be forthcoming.

We can see what happens if this reference frame is lost by looking at teratomas: growths composed of many tissue types that sometimes form spontaneously in the sex organs from germ cells triggered into differentiating at the wrong time or place. Devoid of the correct guidance, the proliferating cells produce random and grotesque bundles of various tissues: muscle, heart, bone, teeth. They are a kind of tumour, some benign and some cancerous, and can appear both in children (including newborn babies) and adults. As the name suggests (from the Greek *tera*: monster), there is something not just strange but rather horrific about this out-of-place yet familiar flesh. Teeth and hair, we can't help feeling, should belong to a person, and not spring unbidden from a disorganized mass of living matter. But what teratomas show us is that normal development is tightly orchestrated as to what and where. If those two coordinates become uncoupled, the result is not just unstructured but uncanny.

Organoids are not as chaotic as that, but neither are they as properly structured as real organs. Mini-brains, for example, display some of sophisticated brain structures, but they don't quite get the shape right. That too creates an unsettling frisson when we appreciate that the neurons are nonetheless functional, sending electrical signals to one another. These are not "thoughts", but they are what thoughts are made from.

Researchers are seeking to supply mini-brains with more of the "environmental cues" they would get in a developing fetus, so that they can become even more brain-like. In a fetus, and indeed even after birth in the growing infant, specific brain regions are shaped by their interactions with other brain regions: they help one another. These processes can be mimicked *in vitro* by bringing together organoids that have been guided to resemble particular brain regions, letting neural connections grow so that they can "talk" to one another in structures called "assembloids" – what you might think of as an improved mini-brain constructed in modular fashion.

The fetal brain also needs to interact with other cell types that don't form part of the brain at all, such as cells that make up some of the blood vessels or immune system of the central nervous system. These too might be cultured separately and then added into an organoid. So-called glial cells – originally thought to be a mere matrix or binder for neurons but now understood to have much more complex roles in the brain* – are likely to be particularly important ingredients of improved brain organoids.

Another key type of signal for the developing brain are those that indicate the body axes – in particular the top–tail (anterior–posterior) and front–back axes that are established early on in embryo growth (see page 64). These signals, we saw, are supplied by cells acting as "organizers", emitting morphogens that establish directions through their concentration gradients. It is precisely these cues that those poignant little bulbs of would-be spinal columns are "looking for". It might be possible to include such directional signals with simple

---

* The word "glia" comes from the Greek for *glue*, a translation of the German word *kitt* given to them by Rudolf Virchow in the 1850s. Virchow and his contemporaries thought these cells were merely a kind of passive matrix – a "nerve-glue" (*nervenkitt*) – binding neurons together. It is now clear that they have a much more active role, fine-tuning the properties of nearby neurons and forging or pruning synaptic connections between them. They have been called a "brain homeostat", maintaining the function and integrity of the neural network.

means: embedding in the organoid beads or tubes of gel infused with the appropriate morphogen proteins. In ways like this, some researchers are already talking about the "next generation" of brain organoids that more closely resemble the real thing.

This isn't because they have any interest in making an organ that can genuinely "think", but because they use brain organoids as the closest alternative to being able to study a real live human brain with invasive and ultimately destructive methods that could not possibly be applied to living people. "You don't need a completely well formed human brain in a dish to study biological questions," Lancaster explains. But if you can improve the resemblance in the right respects, she says, you'll get a better picture of the process in real bodies that you're interested in.

Already Lancaster has used brain organoids to investigate how the size of the human brain is determined: what makes it grow only so large, and no bigger? That's a question elicited by the condition called microcephaly, a growth defect that results in abnormally small brain size. This problem has become a particular focus of research since the onset of the Zika virus pandemic, which began in 2007 in Gabon and Micronesia and spread across the Pacific to South America. For most people, the risks of the virus are minimal: some experience no symptoms at all, while others have only a rash, mild fever and headaches before the virus is cleared from the body. But for pregnant mothers it's another story, because infection can cause microcephaly in the baby. In rare cases, Zika can also cause the condition called Guillain-Barré syndrome, in which neural damage may lead to paralysis and death.

Exactly what the virus does to arrest brain development isn't yet clear, but research on mini-brains might help to understand it. Lancaster, Knoblich and their collaborators have cultured brain organoids from iPSCs made from people with microcephaly and have seen growth anomalies compared to the way healthy organoids develop. Aside from Zika-induced cases, microcephaly has been found to be linked to a mutation of a gene involved in the differentiation of neurons. The researchers have found that if they introduce the normal form of the corresponding protein into brain organoids of microcephalous patients, there's a boost in neuron production.

Lancaster is also interested in what can make brains grow too big – which, contrary to what you might expect, is not a good thing, being linked to neurological disorders like autism. Others are using these mini-brains to study brain conditions such as schizophrenia and epilepsy. Selina Wray is making them to understand the neurodegenerative process in two types of dementia: Alzheimer's disease and frontotemporal dementia. The atrophy of brain tissue may start when two proteins present in nerve cells, called tau and amyloid beta, switch from their normal molecular shape to a misshapen form. These forms stick together in clumps and tangles that accumulate in the brain and cause neurons to die.

By culturing mini-brains from the cells of people with a genetic predisposition to these diseases (who account for about 1 to 5 per cent of all cases) Selina hopes to find out what goes awry with the two proteins as neurons grow. "We hope to see the very earliest disease-associated changes," she says. "That's important when we think about developing treatment." She has found that the tau proteins for the disease samples are different from those in healthy samples.

Studies like this represent the immediate future for organoid research, not just on mini-brains but on other tissue types too: they supply models of organs and tissues in which the progress of human disease can be safely and ethically investigated. It's possible too that human organoids might obviate the need for controversial research on non-human primates. By the same token, organoids could be valuable for testing drugs, allowing their impact on an entire organ to be assessed before moving to clinical trials, without threat to human life from toxicity or unforeseen side-effects. Tumour organoids, which contain a similar diversity of cell types as found in real cancer tumours, have been used to test anti-cancer drugs. One attractive possibility is to use organoids made from iPSCs to find out which drugs might be most effective, and least harmful, to particular individuals. This personalized drug screening might ultimately be automated for high-speed testing by growing organoids on a chip-style device, integrating tiny cell cultures with sensors and delivery systems. It's possible to grow several different organoid types on a single testing platform, leading some researchers to speak – somewhat figuratively, but with unmistakable echoes of Alexis Carrel's visions – of a "body-on-a-chip".

I think we should allow ourselves to feel the strangeness of that image. It needn't involve lurid fantasies of a kind of homunculus mapped out in flat layers of cells as if strapped to a fiendish device and pumped full of drugs. But it would seem nonetheless to exemplify what the German-American biologist Jacques Loeb optimistically envisaged in 1890 as "a technology of living substance". The more accurately and realistically bioengineers can plot out a "schematic human" on a chip, the more useful such a tool could be for medical research.

Other plans for brain organoids are no less dramatic. Swedish palaeontologist Svante Pääbo of the Max Planck Institute for Evolutionary Anthropology in Leipzig is using them to investigate our extinct close relatives the Neanderthals, attempting to figure out what (if anything) made their brains different from those of the early *Homo sapiens* with whom they coexisted.

Neanderthals were a form of human that lived between about 250,000 and 40,000 years ago. They were close enough relatives to make interbreeding possible, which is why most people (if they are not of African descent) have a few per cent of "Neanderthal genes" in their genomes. Pääbo's lab sequenced the Neanderthal genome, patched together from the fossil remains of several individuals, in 2010. Among the differences with the genomes of modern humans, the researchers found distinctively Neanderthal variants in a few genes linked to brain development.

How, and how much, did those genes make Neanderthal brains different from ours? We now know that the stereotype of Neanderthals as dim-witted and "primitive" is wrong. Those people had a rather sophisticated culture that included art and probably rituals, and their brains were in fact bigger than ours (although that doesn't imply they were smarter). All the same, while the reasons for the extinction of Neanderthals aren't clear, it's possible that some cognitive ability of modern humans gave us the advantage. (It's also possible that Neanderthals were simply more susceptible to disease.)

Pääbo hoped to get some clues by using gene-editing methods to transfer some of the Neanderthal genes into human stem cells and growing them into mini-brains. This doesn't by any means produce "Neanderthal mini-brains", as some media reports predictably claimed. But it could show whether those genes alter the development of brains in ways that are reflected in the shapes and structures of the organoids, and which might at least hint at reasons for cognitive differences.

The researchers have indeed found differences in the shapes of these "Neanderthalized" organoids, but the results are hard to interpret given that mini-brains are such a crude approximation to the real thing. At this point, maybe it's enough that the questions can be posed at all.

* * *

The growth of human tissues outside the body holds great potential for making new organs for subsequent transplantation, but certain replacement tissues can't so readily be just stitched into place. In order to function properly, they need to meld intimately with the surrounding tissue as they grow.

Take heart tissue. Heart disease is the main cause of death throughout the world, but heart failure caused by impaired cardiac muscle can't yet be treated effectively by growing and grafting this tissue. Cardiomyocytes – the muscle cells that cause the contraction of heartbeats – need to synchronize their electrical activity throughout the heart. A region of cardiac muscle electrically discon-nected from the rest might go its own way and mess up the rhythm of the heartbeat, potentially causing the condition of arrhythmia that often precedes heart failure. That's why attempts to transplant such tissue grown *in vitro* haven't been successful: the transplant and the rest of the heart don't "talk" to one another in a way that lets them beat in synchrony.

The same is true of regenerated nerve and brain tissue. It would be wonderful if neurons grown in a dish could be used to repair spinal-column injury or brain damage caused by stroke or neuro-degenerative conditions such as Huntington's disease and Parkinson's disease. But for one thing, you'd need the right mix of neurons and other cells: there are hundreds of different types of neuron in the brain, each with its own distinctive shape, pattern of connectivity, and function. What's more, these neurons would have to be wired properly into the existing network. Growing them outside the brain in the right combination and then somehow transplanting them surgically so that they will wire up properly is challenging (although as we'll see later, it is being attempted). But if the neurons were grown in situ, they might get the signals from the surrounding tissues that would direct them into the right cell types and allow them to hook up to one another so that they can send and receive nerve pulses.

One option in these cases is to introduce iPSCs (or ES cells) directly into the body, where their environment might guide them to the required fate. But a key question – to which no one knows the answer, and to which there might not *be* an answer that applies to all cells and tissue types – is whether an adult tissue can supply the guiding cues for stem cells that were present when the tissues first formed during development. There's no guarantee that such cues are still active, or can be reawakened.

Still we can try, and there have been a few attempts so far to develop regenerative therapies by implanting iPSCs from cell culture or making them in situ in the body. In 2014, Shinya Yamanaka collaborated with ophthalmologist Masayo Takahashi to introduce a kind of retinal cell derived from iPSCs into the eye of a patient suffering from age-related macular degeneration. This condition, which causes partial blindness, is due to deterioration of the light-sensitive cells in the retina. It is one of the more promising arenas for stem-cell therapies, since human embryonic stem cells have been shown to develop into retinal cells under the right conditions, and experiments on mice and rats have shown that such cells, grown in culture and surgically transplanted, can detect light and produce nerve signals in response. The big challenge there is actually carrying out the surgery in a delicate organ like the eye.

Yamanaka's trial with injection of iPSC-derived cells had mixed results. Although the progress of the disease seemed to stop, the patient experienced no improvement in her sight. And the researchers had to call off a second trial because of worrisome genetic mutations that appeared in the patient's iPSCs.

Researchers in Japan have also trialled the use of iPSCs cultured into neurons to treat neurodegeneration. Takayuki Kikuchi, a neurosurgeon at Kyoto University hospital, has implanted the precursor cells of neurons, grown from iPSCs taken from a donor, into the brain of a man in his fifties suffering from Parkinson's disease. This condition, for which there is currently no known cure, arises from the death of neurons responsible for releasing the neurotransmitter dopamine, which is involved in movement. Implantation of iPSCs themselves directly into the brain has been shown to improve the symptoms for macaque monkeys afflicted with a Parkinson's-like condition, making them more mobile, without any apparent adverse side-effects.

The iPSCs in the human trial could produce dopamine, and Kikuchi's team implanted 2.4 million of them in various sites in the patient's brain known to be sites of dopamine activity. The initial outcome seems promising – the patient appeared to remain well – and if things stay that way then the researchers plan to implant more of the cells later. They think it is possible that the treatment could become generally available by around 2023. Another team of Japanese scientists, led by Hideyuki Okano at Keio University, is planning a trial to inject iPSC-derived neuron precursor cells into the sites of damage for patients with spinal-column injuries.

These efforts use pre-cultured iPSCs from a donor, rather than having to go through a months-long process of extracting and growing them from the patient. Yamanaka plans to culture "standardized" iPSC lines that can be matched to the immune systems of patients – a bit like blood-group matching for transfusions – so that the risks of rejection are reduced with only mild doses of immune-suppressing drugs.

Another possibility is to reprogramme somatic cells to iPSCs directly within the body at the site of injury or damage. Researchers at Indiana University have tried that strategy to treat brain injury. When the brain is damaged in the outermost layers (the cortex), it starts producing glial cells. It's not known if this is a response that causes more deterioration, or if on the contrary it is a kind of repair or damage-limitation strategy. Either way, the new glia end up as scar tissue in the brain, useless for any cognitive function. But the Indiana researchers figured that these glia might be usefully turned instead into new neurons to replace the damaged ones.

They injected viruses encoding Yamanaka's four-component iPSC cocktail directly into the brains of mice that had received a cortical injury, and confirmed that it switched some (albeit relatively few) of the newly proliferating glia into an induced stem-cell state. These iPSCs then went on to develop into functioning neurons. It's still a big step from here to restoring anything like the brain functions that such injury impairs, however. And while introducing stem cells directly into the body, whether by transplantation or in situ reprogramming, can assist the integration of the new tissue with that which is already present, it also carries a grave risk, for the fate of such cells can't be guaranteed. In the body, they have a tendency to develop not into healthy tissue but into cancer cells.

Fortunately, there's another alternative.

* * *

If differentiated cells can be switched back to stem cells by a suitable mixture of transcription factors, perhaps their fate can be rewritten in other ways too? Might somatic cells of one type be transformed *directly* into another – skin into heart or muscle, say – in a feat of biotechnological transmutation?

Yes indeed. It seems again to be largely a matter of finding a package of transcription factors that can persuade a cell it has a different nature. The creation of iPSCs is, in this view, just one particular instance of a more general technology for reprogramming cells. In an analogy that reinforces the notion of cell lineages attaining their fates in a process of computation, cell biologist Deepak Srivastava, president of the Gladstone Institutes for biomedical research in San Francisco, says that with these transformations "you're basically hacking the cell's code".

The first attempts to switch cells directly from one "mature" fate to another were in fact made well before Yamanaka had demonstrated that somatic cells could be turned into stem cells. In 1987, researchers at the Hutchinson Cancer Research Center in Seattle reported that adding the right genes to *in vitro* cultures of fibroblast cells could turn them into muscle-forming cells (myoblasts).

But the potential of this approach has only become truly apparent since Yamanaka's work, which revealed just how unexpectedly malleable cell states are. For example, in 2010 Srivastava and his colleagues made mouse cardiomyocytes from heart fibroblasts by treating them with three transcription factors. With a few subsequent improvements, the reprogrammed cells showed the coordinated "beating" behaviour of real cardiac muscle. And in 2013, the same transformation was demonstrated – still in the petri dish, mind you – for human cells.

Rather than using transcription factors to reprogramme human cardiac cells, it can be done using wholly synthetic molecules as the switching agents, which seem able to mimic the effects of the natural proteins. Typically these molecules are found by trial and error: researchers create vast libraries of small molecules at random, with somewhat related shapes and compositions, and then test them in cell cultures to see if any have the desired effect.

Perhaps it's not so surprising that human-made chemicals can do something like this. After all, many drugs exert their effect by imitating the roles of natural biochemicals in the body. All the same, the idea that cell reprogramming can be achieved with molecules we can make in the lab in much the same way as we devise, say, new pigments or plastics, opens up new vistas. Might such synthetic molecules be found that can alter the fates of cells in ways not found in nature? Might they even create entirely new cell states – new types of flesh, one might say? We don't know, but it's possible.

It's not just heart tissue that can be generated this way. In 2010, for example, researchers at Stanford University in California showed that three transcription factors would convert mouse fibroblasts directly to neurons. A year later, it was done in human cells. In some cases, just a single transcription-factor gene (Sox2) is enough to do the job.* What this particular piece of cellular conjuring amounts to is that it accomplishes what was done to my chunk of arm-tissue, but without the intermediary stem cells.

The ultimate aim is to do this sort of thing directly in the body. Mostly, such transformations have been trialled on mice. In 2008, Douglas Melton of Harvard University and his colleagues used viruses to deliver three transcription factors to live mice that turned certain cells in the pancreas, called exocrine cells, into the ones that make insulin in that organ, called beta cells. As is the case for many such experiments on direct reprogramming of cells *in vivo*, the switching here is a relatively modest change, interconverting cells that are already closely related in a lineage.

Pancreas and liver cells also share descent from the same progenitor cells during embryo growth, so researchers have figured it might not take too much of a nudge to produce pancreatic beta cells in the liver. (To fulfil their role, it really doesn't much matter where beta cells are in the body, so long as they make insulin.) Jonathan Slack and his co-workers at the University of Minnesota achieved that switch in mice and found that it could alleviate the symptoms of diabetes in these animals.

Curiously, the three genes that turn pancreatic exocrine cells into beta cells in living mice *won't* induce the same change if added to the same cells in a petri dish. This suggests that signals from surrounding tissues can make direct reprogramming more effective and efficient when done inside the body. That's generally the case,

---

* You might remember that Sox2 is one of the transcription factors in Yamanaka's cocktail for turning fibroblasts into iPSCs. So why is it here making neurons? This illustrates again the subtlety within the black box of genetic circuitry. Sox2 is evidently not by any means a "stem-cell-making" gene – it has a more general but hard-to-define influence on cell development that depends on the company it keeps.

and it is rather convenient, although no one fully understands why it's so.

In humans, severe type 1 diabetes is commonly treated by transplanting insulin-making clusters of pancreatic cells called islets. But there is a chronic shortage of donor islets, and immune rejection of the transplant complicates the procedure. So direct reprogramming of a patient's own cells into pancreatic cells in the body looks like an attractive alternative. It won't be easy though. One study on *in vivo* reprogramming of pancreatic cells in macaque monkeys – a much better animal model for humans – elicited only transient production of insulin: it came and then went again. What's more, those results suggested that large doses of the switching agents might be needed, which is never a happy circumstance when administering drugs.

Few faculties are unimpaired by the ageing process, but among the hardest losses to bear are sight and hearing. Of course, illness and accident can damage both of these capacities much earlier in life too, and some people lack them from birth. It's not surprising that giving eyesight to the blind has long been a power granted to saints and saviours, the miraculous act of healing serving as a universal spiritual metaphor.

In many cases, the requisite medical miracle is repair of damaged or deteriorated retinal tissue. Humans can't regenerate their retinas – but zebrafish can. If these animals suffer retinal damage, the activity of a transcription-factor gene called Ascl1 is boosted in a type of retinal cell called Müller glia, which transforms them into light-sensitive neurons: this is in effect a kind of natural cell reprogramming. Mammals like us have versions of this gene too, but it has a rather different role: prompting embryonic cells to become neurons during early development. That's a one-time process: our bodies can't switch our Ascl1 gene back on in the retina if it is injured later in life. But researchers at the University of Washington in Seattle wondered whether introducing an active Ascl1 gene artificially in mammals would have a similar effect to its activation in zebrafish, switching Müller glia to neurons. They found that in young mice a dose of Ascl1 will do just that, although the effect fades. But if the researchers boosted another (enzyme-encoding) gene at the same time, the glial cells of adult mice would develop into neurons that respond to light. That's far from a cure for blindness, but it's a hopeful start.

What about reversing hearing loss? Fish, birds and frogs can regain some auditory ability if the sound-sensitive "hair cells" of their ears are damaged. (These aren't cells that grow into hairs, but cells with a hair-like shape which activate auditory nerves in response to the vibrations caused by sound.) A key transcription factor involved in hair-cell growth in embryos is called Atoh1, and adding this gene will reprogramme cells in rats to make hair cells (the conversion is more efficient with a few additional ingredients). To restore hearing this way, though, the new hair cells must get properly wired into the auditory neural network, and this hasn't yet been demonstrated.

Reprogramming of cells to make neurons directly in the body could have particularly dramatic medical implications: to repair paralysing spinal-column injury, say, and maybe even brain damage or deterioration. Chun-Li Zhang at the University of Texas Southwestern Medical Center in Dallas and his co-workers have shown that glial cells in the brain can be transformed in adult mice to neuron-forming cells called neuroblasts, from which healthy neurons could potentially be generated. Again it's no more than a promising start. Even if neurons are formed directly *in vivo*, it remains to be seen whether they will properly integrate with and "talk to" their neighbours.

And what of the heart, that most symbolic, vital and vulnerable of organs? In 2012, Srivastava's team at the Gladstone Institutes succeeded in reprogramming cardiofibroblasts (non-beating heart connective tissue) into cardiomyocytes by injecting the same cocktail of three transcription factors that were known to effect direct conversion *in vitro*. Again, the reprogramming was more efficient *in vivo*. At the time of writing, the San Francisco startup company Tenaya Therapeutics, which is affiliated with the Gladstone, is planning to start clinical trials for *in vivo* reprogramming of heart cells, involving people who already have serious, high-risk conditions causing arrhythmia and who currently rely on prosthetic implants to avert it.

Changing the very nature of our flesh inside the body sounds extraordinary, but it could also be risky. In theory, there is a danger that the reprogramming molecules or genes in the body might run amok, drifting away from the place where they are supposed to work and switching other cells elsewhere. But that doesn't seem to cause problems in the animal experiments conducted so far. Switching a cell's fate is a big deal: it takes a hefty and very specific push to move a cell from one mature state to another, and the conditions have to be just right. So a bit of stray reprogramming agent is unlikely to be powerful enough to have any effect on cells very different or distant from its target.

As well as commandeering existing cells for new functions, there is another alternative for *in vivo* reprogramming that is in some ways even more suggestive and dramatic: to make existing, specialized cells start growing again, just as they did while we developed *in utero*. Your heart or kidney can't repair itself because it has simply stopped growing. There's a strict end point to tissue growth, which is why we don't continue getting bigger throughout our life. That shutdown is what makes us unable to grow a new limb.

But what if the growth process could be re-started? Researchers are looking for ways to do that, typically by seeking the genes that are active in proliferating cells in embryonic and fetal mice in the hope that boosting their expression in adult cells could get the process underway again. Cell proliferation is controlled by the cell cycle: the sequence of events that cells undergo when they replicate and divide. It's not generally something you want to tamper with, since, as we saw earlier, breakdown of the cell cycle is the common road to cancer.

Nonetheless, with care, transcription factors involved in the cell cycle can indeed be used to bring adult cells back into something like the state they were in as the body was still growing, so that they can produce fresh tissue in what amounts to a form of rejuvenation. The process has in fact been known for some years, but the challenge is to make it happen efficiently enough to make a difference. Recently Srivastava and his colleagues have been able to improve the efficiency with which new cardiomyocytes will grow in adult mice. When they injected certain transcription-factor genes into adult mice that had damaged heart muscle, before scar tissue began to develop, they found that the heart function was improved. Instead of making non-functional scarring, these mice were apparently able to generate fresh heart muscle, something that would never normally happen for adult cells. The same group of factors will reawaken cell division in human heart cells cultured outside the body, though a lot of careful testing remains to be done before we know if it is safe enough to try on patients in the clinic.

* * *

Now that the plasticity of our cells has been revealed, we can see that evidence of this fact had existed, but been overlooked, for a long time. Arguably, the first demonstration of *in vivo* cell reprogramming happened in 1891, when the zoologist V. L. Colluci showed that newts that have the lenses of their eyes surgically removed can regenerate them. It was found a few years later that the cells of the new lens were previously pigment-containing cells in the iris, which were able to "de-differentiate" before taking on a different fate.

Newts and other amphibians are, as we've seen, rather special among vertebrate organisms in their ability to regenerate tissues. But it is starting to seem now as though cell reprogramming, including reversals to a stem-cell-like state, is a normal feature of nature. Cell biologists are currently debating whether the adult stem cells that replenish specific tissues are truly stem cells all along, or whether they might be generated on demand by the de-differentiation of mature somatic cells. Blood and muscle do seem to maintain populations of dedicated stem cells, but other tissues acquire what are called "facultative stem cells": specialized cells that revert to stem cells when the tissue is injured and needs to repair itself. Our gut walls contain dedicated stem cells, but if these are depleted for some reason, or if the gut is injured and new cells are needed fast, some of the ordinary epithelial cells of the lining can turn back into stem cells and repopulate the gut wall. Facultative stem cells have also been found in the lungs and, at least in mice, in the liver.

Thus, as Yamanaka and fellow cell biologist Alejandro Sánchez Alvarado have written:

> It appears that, in nature, for many cells and in many organisms, the differentiated state is not terminal but is instead [merely] stable and that, under varied environmental conditions such as injury and disease or even the natural process of ageing, such a state can be reprogrammed.

This makes perfect sense from an evolutionary perspective. If cells can in principle retract their lineage back up the valleys of Waddington's epigenetic landscape (permit me here to revive that simplistic but useful description) – or if they can hop across valleys

– then why on earth wouldn't natural selection leave that option open for dealing with traumatic times? Such transformations have to be carefully managed, of course, to avoid the risk of tumours. But cells are clever. That cleverness depends on being responsive to their neighbours and their environment, attuning their state to the demands of the moment.

As understanding of cell plasticity, both natural and induced, advances almost weekly, there is not the slightest prospect that I can future-proof a discussion of this field here – it is moving faster than the still somewhat stately machinery of publishing. On balance, I'm glad of that. Sure, we might eventually find fatal flaws in these plans to rewrite our flesh – but I think it is more likely that the ingenuity and tenacity on display so far will let us navigate the obstacles, devise new strategies, and some time soon begin bringing relief to humankind's estate. I am old enough to feel the pressing need of techniques that can patch up our ailing bodies but, if I am lucky, young enough to benefit from this revolution.

That's a strong and overused word, but I believe it earns its place here – not so much because of the medical advances on the horizon but because they rest on a conceptual transformation that has already taken place. There is every reason now to claim that any part of our body might be changed into any other part, if we can find the right dials within our cells. We can look at nature, where the salamander regrows a leg or the fish repairs an eye, and it is not a vain hope that we can learn those tricks – that we retain enough vestigial kinship with these marvellous creatures to emulate the genius of their tissues.

CHAPTER 5

THE SPARE PARTS FACTORY

MAKING TISSUES AND ORGANS
FROM REPROGRAMMED CELLS

Growing entire human organs was always a dream of the tissue culturers. While Alexis Carrel laboured with his protégé Charles Lindbergh to keep human organs alive outside the body by perfusing them with blood plasma in artificial containers, the ultimate prize was "the culture of whole organs" – the title of an article that the two men published in *Science* in 1935. That paper didn't report anything of the kind, but when Carrel saw new tissue growing on a perfused cat ovary, he concluded that organs should be able to make themselves from their constituent cells.

"Isolated cells," he wrote in his book *Man, the Unknown*, also published in 1935, "have the singular power of reproducing, without direction or purpose, the edifices characterizing each organ." These organs, he added, are "engendered by cells which, to all appearances, have a knowledge of the future edifice, and synthesize from substances contained in blood plasma the building material and even the workers." It seemed almost magical, and even the arch-rationalist Carrel allowed himself a little flourish, saying that "an organ develops by means such as those attributed to fairies in the tales told to children in bygone times." This kind of magic, he hoped, would lead to another: to immortality, enabled by a constant renewal of organs made in the laboratory.

Thomas Strangeways and Honor Fell at the Cambridge Research Laboratory found that tissues cut from embryos after the cells had differentiated and then cultured *in vitro* would continue to grow as the same cell type: cells of eyes or bones, say. This was an early form of what today is known as *tissue engineering* – a striking lexical amalgam of the biological and the industrial, which seems to imply that growing a human (or parts thereof) is at root an engineering problem. In 1926, Strangeways proclaimed that "somatic cells do not require the control of the organism as a whole in order to build up the specific tissues for which they were set apart in life." There is no clearer statement of the philosophy that motivates, and receives support from, work today on mini-brains and other organoids.

In the late 1910s, the Scottish biologist David Thomson, working in London, travelled to learn from Carrel at the Rockefeller before returning to England to grow nascent organs of chick embryos excised and cultured *in vitro*. He found that they retained their anatomical structure and argued that this was because a membrane around the growing organ prevented uncontrolled proliferation of cells at the edges. But as Thomson rightly discerned, there was a limit to how big such "artificial organs" could get, because the nutrient would at some stage fail to reach the innermost cells and they would die. He recognized that to overcome this limitation demanded "some means [of] artificial circulation": the cells needed a blood supply.

Culturing and sustaining organs *in vitro* was the enabling technology that led biologist J. B. S. Haldane to imagine the production of humans wholly outside the body. As he explained in a lecture of 1923 in Cambridge that he developed the following year into a book called *Daedalus, or Science & the Future*:

We can take an ovary from a woman, and keep it growing in a suitable fluid for as long as twenty years, producing a fresh ovum each month, of which 90 per cent can be fertilized, and the embryos grown successfully for nine months, and then brought out into the air.

Haldane's speculations were the prelude to IVF and assisted reproduction technologies, as we will see in the next chapter. But such visions of making humans have never been distinct from work on organ and tissue culture. Haldane himself drew confidence from the results coming out of Strangeways's lab, and *Nature*'s review of *Daedalus* commented that the book didn't seem far-fetched "if what has already been done with tissue culture is remembered".

The growth of parts of organs like eyes and bones *in vitro* at Strangeways led some observers to believe the lab was on the point of making entire organisms piece by piece. "Already other parts of the human body have been grown in test-tubes," wrote Norah Burke in *Tit-Bits* in 1938. "Remember that chicken heart that went on growing and growing . . . It may be that the human species will produce test-tube babies or full-grown human beings, entirely chemically."

It wasn't just wild fantasy, given what some scientists were already claiming. By 1959, the French biologist Jean Rostand would assert (with little justification, it must be said) that "it is now possible to construct, as it were, artificial organs, complete with heart and lungs . . . it has been done by Carrel and Lindbergh . . . and many others – whose characteristics approach nearer and nearer to the natural." With such ideas in the air, you could hardly blame the likes of Burke for getting overexcited.

* * *

But could cells really do it all, growing into an entire organ outside the body? Some researchers looked askance at cells cultured *in* *vitro*, considering them to be an undisciplined mass: as the Russian histologist Alexander Maximow put it in 1925, "a crowd of various cells without any regular arrangement".* There was a suspicion that the whole organism, or at least some substantial portion of it, might be needed to instil the order evident in real tissues and organs.

In any event, despite the hyperbolic and sometimes ominous press reports that accompanied early work on tissue culture at Strangeways and Carrel's lab, human tissue engineering never took off at that time. There were successes with other mammals and higher organisms, but researchers found that normal human cells were peculiarly difficult to sustain. By the 1960s, efforts to grow human organs and tissues from the cells up had all but disappeared.

A few researchers persisted. In the 1970s, there were efforts to make artificial skin for covering the wounds of burns victims to aid healing and prevent infection. John Burke of the Massachusetts General Hospital developed thin sheets, made mostly from the natural protein collagen, that would support the migration and proliferation of skin cells when placed over the wound. Thus Burke's material acted not just as an emergency covering but as a kind of biodegradable scaffold to support skin growth *in vivo*. This technique has now become rather routine: several commercial polymer products exist, typically made from cow collagen, on which skin cells (keratinocytes and fibroblasts) can grow in either *in vitro* culture or directly on a wound. Materials like this have been used, for example, to treat otherwise non-healing diabetic foot ulcers that would previously have required amputation. Artificial skin has also been grown from umbilical-cord stem cells and stored as ready-made sheets for surgical use. Although there remains the problem

---

* Although Maximow was working at that time at the University of Chicago, medical historian Duncan Wilson points out that his disquiet about the anarchy of anti-Bolshevik uprisings in his homeland around this time might account for the quasi-political overtones in this choice of words.

of skin grown from one individual's cells being rejected when grafted onto another person, it can be reduced by immune matching of donor and recipient (see page 135).

"Synthetic skin" grown from stem cells exemplifies another motivation for growing tissues and primitive organ-like structures outside the body: not for surgical use but for testing drugs and other pharmaceutical products. A good enough mimic could reveal potential side-effects – for example, the propensity for a medication to provoke inflammation and skin irritation or to prove toxic. Such testing has traditionally relied on animal experiments, which are not only ethically controversial but also sometimes of questionable relevance to humans. *In vitro* cultured skin has already been put to use in this way, and tissue cultures that supply approximations to the human kidney, liver or brain may likewise help to sift through drug candidates for those that work safely and effectively – perhaps even in a way that is personalized to the patient, since drugs don't always produce the same effects, for better or worse, in every individual. That approach remains to be validated, however – cell biologist Marta Shahbazi warns that "we may find ourselves curing organoids but not curing people."

The skin is our largest organ, but also in many ways the simplest. Making other artificial organs is more challenging. In the mid-1980s, paediatric surgeon Joseph Vacanti, also working at the Massachusetts General Hospital, became frustrated that his efforts to save children's lives through transplants were hindered by a shortage of organs. "I watched in agony and completely helpless as several children faded into coma or haemorrhaged to death," Vacanti recalls. One of the few options available for obtaining livers for transplanting to children was to trim adult livers to size – but that crude approach was fraught with difficulty. "It occurred to me that if we could build liver tissue, we could transplant on demand," Vacanti recalls. But how?

Vacanti had witnessed Burke's efforts to grow skin on polymer supports during his stints on the burns ward early in his training. But a liver is a very different sort of tissue to skin – for one thing, like most organs it needs a system of blood vessels to keep the cells alive.* Vacanti turned to his colleague Robert Langer, a chemical engineer at the Massachusetts Institute of Technology, who had studied ways to control the development of blood vasculature (the process called angiogenesis) as a possible means of shutting down the growth of cancer tumours.

Together, Vacanti and Langer attempted to culture liver cells on an artificial scaffold, using polymer materials similar to the earlier work that were already approved for human use. One way of providing the cultured tissue with a blood supply was to simply build it in: to lace the polymer support with a network of tiny channels, which would be seeded with the cells (called endothelial cells) that make the walls of blood vessels. Langer and his co-workers have made a kind of artificial liver by stacking alternating layers of polymer sheets patterned in this way, one layer seeded with the synthetic blood vessels and the next with liver cells. In this sandwich arrangement, no liver cell is ever too far from a blood supply. Structures like this aren't intended as long-term artificial livers, but the researchers hope that they might be implanted to sustain patients with liver failure who are awaiting transplants. One big challenge is to connect the body's own blood-vessel network to that of such an artificial organ. A promising approach is to seed the scaffolding material with proteins called growth factors that stimulate the body to grow vascular networks, so that natural blood vessels might infiltrate it.

Laryngologist Martin Birchall of University College London was,

---

* Any human tissue thicker than about a few tenths of a millimetre needs to have a vascular system (blood vessels) to keep the cells well supplied with oxygen and nutrients.

like Vacanti, frustrated by the limitations of conventional transplant surgery. While doing head and neck cancer operations, he says "it became clear that, even with the very best of modern techniques, we were still a long way from being able to restore function and quality of life to patients who had had major treatment to mouth, tongue, larynx, pharynx and oesophagus." Birchall felt that this impairment in quality of life for his patients could be dehumanizing. "I felt there must be a better way to go," he says.

For the kind of airway surgery that Birchall carried out, replacement tissue structures would ideally be personalized in shape to fit the patient. This could be done by shaping a polymer scaffold which is then colonized by cells *in vitro* before being surgically implanted. In the mid-2000s, Birchall developed a tissue-engineered windpipe this way that could be transplanted to pigs. And in 2008, he and his Spanish collaborators got permission to use the technology on a young Spanish woman who had suffered life-threatening damage to her windpipe after a tuberculosis infection. The researchers seeded a scaffold with the patient's own stem cells, taken from her bone marrow, to avoid problems of immune rejection. "Considering how little we knew at the time, it worked amazingly well," he says. The woman is still alive and well today.

Usually such stem cells need signals to trigger proliferation and to guide them towards the required cell fate. For example, so-called mesenchymal stem cells taken from muscle, bone marrow or fat will adopt a different fate depending on how stiff the material is within which they are growing: they develop into the cell type of the tissue with the closest match in stiffness. It's as if the stem cells give a little tug on their surrounding environment to figure out what they should become. So the differentiation of these cells can be guided by mechanical cues in their surroundings. Alternatively, biochemical agents such as transcription factors can be added to determine the cells' fate.

Tissue engineering with adult stem cells is still a relatively new technique, but it holds tremendous promise. The several different cell types that might be present in a single organ, such as vascular cells, bile and liver cells in a liver, typically share a common lineage and so might all be made from the same stem cells if given the right cues. An alternative that avoids the difficult business of harvesting adult stem cells from a patient is to use induced pluri- potent stem cells cultured from, say, their skin. Langer and Vacanti think that iPSCs could become "the ideal material for building tissue constructs" – so long as their tendency to develop instead into tumours in the body can be kept in check. Embryonic stem cells can be used too, but then you're also up against the problem of immune rejection, necessitating doses of immune-suppressing drugs.

Today's architects of artificial organs are upbeat about the pros- pects. In 2002, Robert Lanza of the American tissue-engineering company Advanced Cell Technologies made the bold prophesy that

> If this research [on stem cells] is allowed to proceed, by the time we grow old, this will be a routine thing. You'll just go and get a skin cell removed at the doctor's office, and they'll give you back a new organ or some new tissue – a new liver; a new kidney – and you'll be fixed. And it's not science fiction. This is very, very real.

"With every year I see more hope even for the most difficult prob- lems of tissue regeneration, such as brain repair after strokes or spinal-column damage," says Vacanti. But big problems remain even for relatively simple tissues. For example, although Vacanti and others engineered artificial cartilage in the mid-1990s – which, like skin, doesn't need an extensive network of blood vessels – the materials still haven't been used on humans because the wound- healing process after transplantation is complicated. Synthetic cartilage tends to be re-absorbed by the body over time, causing the tissue to deform. Cells have their own agendas, which we don't yet fully understand, let alone know how to control. "This is difficult science," Vacanti admits.

*   *   *

Another option for culturing cells into artificial organs with a predefined shape is to dispense with synthetic polymer scaffolds and use in their place the "skeleton" of an actual donor organ instead. The cells of organs are bound together by a robust network called the extracellular matrix, a web made from a variety of biomolecules that the cells secrete. In animal tissues, these are typically sugar-based polymers (polysaccharides) and fibre-forming proteins such as collagen and the stretchy elastin. Cells have molecules on their surface that stick securely to these matrix components. (It's precisely because collagen is a component of the extracellular matrix that it makes such a good synthetic scaffold material.)

The idea is to use detergents and enzymes to wash away all the native cells in the donor organ, leaving just the "decellularized" matrix on which the recipient's cells can then be grown. Since none of the donor cells remain, animal organs such as pig hearts can also provide decellularized supports for growing human organs.

For simple soft tissues like skin, decellularization is already used to make commercially available products, for example using the skin (dermis) and intestine of pigs and cows, and indeed of humans. Martin Birchall has experimented with decellularized trachea scaffolds taken from both pigs and humans; in fact, his 2008 operation on the Spanish patient used such a segment of trachea taken from a deceased 51-year-old woman. For complex organs, studies haven't yet advanced beyond animal experiments. Lungs, kidneys and hearts have all been grown on decellularized scaffolds from rats, but the outcomes from subsequent transplants have been mixed.

The rat kidneys produced a urine-like liquid, for example, but the lung quickly filled with fluid.

In 2013 a team at the University of Pittsburgh School of Medicine reported the growth of a human "mini-heart" on a mouse scaffold. They seeded the decellularized mouse heart with human cells grown from iPSCs into the progenitor cells of cardiovascular tissue. The cells not only spread throughout the scaffold but differentiated into the specialized types of the heart: cardiomyocytes, other muscle cells, and vasculature-forming endothelial cells. The researchers perfused the artificial organ for 20 days with a culture medium that included growth factors, at which point it started showing spontaneous contractions at a rate of 40 to 50 per minute: it was, after a fashion, beating. The artificial heart also responded to drugs known to affect the beating behaviour in humans.

A beating, mouse-sized human heart in a dish? Well, sort of. Just because a heart contracts doesn't mean it works properly. All the same, the result bodes well for making human-sized hearts from the corresponding decellularized organs from, say, pigs.

* * *

Tissue engineering treats living flesh as a material for moulding, shaping, transforming. If ever there was a technology that showed just how far we have come with that philosophy, it is 3D bioprinting. Here, cells themselves comprise the "ink" that is dispensed through a fine nozzle just like the coloured inks of your home printer, to build up complex three-dimensional shapes layer by layer.

3D printing is already transforming the art of manufacturing in general. To make objects ranging from machine components to artistic sculptures, the print heads typically squirt out resins or powders of metal, ceramics or plaster that can be treated to weld or set them into robust structures. Under computer control, these systems can produce intricate forms of almost any shape, from pottery to engine parts. They have even been used to sculpt food, producing ornate creations in pasta or chocolate. Textiles too can now be printed rather than spun and woven. As 3D printers become cheaper, companies and even individuals can produce artefacts on demand without having to order them from suppliers, just by downloading the appropriate printing instructions. Some futurologists foresee a time when "shopping" will mean pressing the Print button – although we should hesitate to equate technological possibility with commercial and socioeconomic viability.

Why not, then, make body parts in this way too? In one of the first clinical applications of 3D bioprinting, in 2014 a patient admitted to Morriston Hospital in Swansea, Wales, after a motorcycle accident underwent surgery that used tailor-made, printed titanium components to hold together his damaged facial bones. Bespoke metal implants or biodegradable polymer scaffolds like this, designed to fit a particular patient's body, could become widespread; they are already used in craniofacial surgery for shaping bone-substitute materials. 3D printers can construct the most convoluted structures or reproduce exactly the shape of the organ or tissue that needs replacing. You can scan a patient's anatomy before an operation using the technique of computerized tomography – an advanced form of X-ray scanning – and then use 3D printing to produce a material that is exactly the right size and shape. In this way, researchers at the University of Michigan have tailored a purely synthetic polymer sleeve to prevent a malformed section of an infant's trachea from collapsing and blocking his airway. Larynx (voicebox) implants too have to be rather precisely shaped to fit the patient, and Birchall forecasts that they will one day be made from 3D printed biodegradable scaffolds seeded with the patient's iPSCs.

The possibilities become even more dramatic when the inkjets dispense biological tissue itself: when we start printing flesh. Here the inks consist of clumps of living cells, typically protected from the trauma of high-speed passage through a nozzle and impact on a surface by being encased within droplets of a soft, biocompatible polymer or gel material. This technology is in its infancy, but it's clear already that cells can survive the process and be assembled into complex shapes. 3D inkjet printing has been tested on animals to speed up wound healing by spraying cells that make skin and cartilage directly onto the site of injury. Flat sheets of tissue are relatively easy to make, and tubular structures like blood vessels and the trachea are feasible too. Solid 3D organs like livers and hearts would be much more challenging to "print", and no one imagines that will happen soon.

Another, cheaper bioprinting method is extrusion – like squeezing toothpaste out of a tube, but with a much finer nozzle. Here the "paste" is usually again a soft, biocompatible polymer infused with cells, which is "written" into the desired pattern a layer at a time. Not all cells tend to survive the squeezing, but usually more than half of them do.

3D bioprinting could supply patterned vascular networks for organs and tissues grown *in vitro*. A team at Harvard University led by Jennifer Lewis has grown tissues that contain three-dimensional grid-like networks of vessels. First they print the grid using a polymer ink that can be washed away later, thus acting like a removable wax mould in a conventional casting process. Then they print the cells (coated with gel) around this network, fusing the gel droplets together to make a robust material. Extracting the "vascular" ink leaves behind open channels running through the artificial tissue. The researchers infuse these passageways with epithelial cells, which form an impermeable, blood-vessel-like coating on the walls. In this way, the team have made tissues of fibroblasts more than a centimetre thick, in which the cells can be kept alive for more than six weeks by blood flowing through the artificial network. By delivering the right transcription factors into the growth medium, they induced the cells to differentiate into a bone-forming variety and could see the beginnings of bone growth. It's a first step towards making new bones, laced with blood vessels, that could exactly copy the shape of ones that have been irreparably damaged or deteriorated.

Vladimir Mironov, who heads a Russian startup company called 3D Bioprinting Solutions, is convinced that bioprinted human organs are on the way. His company is currently aiming to make a mouse thyroid gland this way – and in the longer term, a human kidney. Mironov has no qualms about extrapolating the vision all the way to what he calls "the realization of Pygmalion's dream": printing an entire functional human body. (The legendary sculptor Pygmalion made a statue of a woman so beautiful that he fell in love with her, whereupon a blessing from the goddess Aphrodite brought the statue to life when he kissed it.) The self-organizing cells will take care of the details themselves, Mironov says; all the printer need do is put them in more or less the right place with the right density.

I'm not sure we need take this notion too literally. It would have been grist for the mill of *Amazing Stories* – "The Man Who Printed His Wife" would be just the kind of unexamined metaphor in which that barometer of techno-cultural change revelled. But a printing process for humans hardly addresses any burning clinical or social need.

Cynics might therefore see Mironov's suggestion as sheer hype. But we might more generously regard it as a provocative thought experiment. For the point about 3D printing is that you can make any shape you like, and it's not obvious that a 3D-printed organ would have to recapitulate the exact shape or cell organization of the natural variety in order to work properly – Jennifer Lewis's printed vasculature does not, for example. A simplified, idealized structure might be perfectly adequate, perhaps with greater geometrical regularity that makes it easier to manufacture. No one knows how amenable tissues and organs might be to redesign.

You might then ask the same questions of an entire bioprinted body. Perhaps this too needn't be shaped exactly like a human body, or even have a recognizably humanoid form at all. Again, rest assured that this is purely hypothetical; no bizarre bioprinted organism is going to twitch, groan, and sit up on the manufacturing platform any time soon. But the scenario can be posed, and it shows that the technologies of cell transformation and construction are allowing us now to formulate and even to begin investigating profound and disconcerting questions in biology. What are the limits of what we can, and should, grow? What are the design constraints of a human body? What qualifies as a human?

* * *

The instincts of an engineer are to build. Those of a biologist are to grow.

So far, making artificial organs from cultured cells involves a bit of both. Whether by using shaped or decellularized scaffolds, or by positioning cells using 3D printers, the idea is to guide the culture towards the correct structure for a tissue or organ by artificial means. Neither approach, though, truly reflects what happens in the body. Here, as we saw, tissues are shaped by other tissues through a complicated dialogue between cells involving chemical and mechanical cues, movement, adhesion, self-organization.

In the early days of tissue culture, there was much debate about whether or not the whole organism is needed to impose order on proliferating cells. Research on organoids has now delivered an answer, and it is the one very familiar to biologists: yes and no. Cells of a particular tissue type can do an awful lot of the arranging via their mutual interactions, but they don't get it quite right without the correct signals from their environment within the whole growing organism. The more you can make an organoid "think" it is part of an embryo, the more it is likely to resemble the real thing. As with children, you can force cells to be a certain way but it might be better to let them find their own way there, with just enough gentle instruction.

And the best instructor of all is a (developing) body.

Where, though, can you get a body that will patiently and expertly tell cells how to grow into organs? Well, it needn't be a human one. We can grow human organs inside other animals.

Your instincts might rebel here, and rightly so. We learn in basic biology that species are fundamentally incompatible, made distinct by the fact that they can't interbreed. In truth, it's a little more complicated than that, for some closely related species *can* have progeny – that's how mules are made from the crossbreeding of a horse and a donkey. Indeed, cross-species breeding is more widely possible in principle than you might imagine: tigers and lions, for example, can produce either a tigon (male tiger/female lion) or liger (vice versa). Neither are such hybrids necessarily sterile: the mule is, but tigons and ligers are not. Such crossbreeds are rare in nature more because of instinctive mating habits and separation of habitats than because they are biologically impossible.

All the same, species that can interbreed to make hybrids must be very closely related in evolutionary terms: a tiger can't mate with a horse. So how on earth could a human organ grow in another animal?

Yet it can. Take, for example, the work reported in 2013 by Takanori Takebe of the Yokohama City University Graduate School of Medicine in Japan and his co-workers. They made human iPSCs and guided them into becoming precursor cells for making liver tissue, called hepatic endoderm cells. Then they cultured these cells in a mixture with endothelial cells taken from umbilical tissue and mesenchymal stem cells. Contrary to the prevailing belief that cells in culture can't mimic the complex interactions that lead to organ formation, this mixture of cell types organized itself into structures that look like the early form of an embryonic liver, called liver buds. In a normal human embryo, such structures appear around the third or fourth week of gestation.

But the organoids could get no further *in vitro* because of the restrictions imposed by lack of a vascular network. To get around this problem, the Japanese team transplanted their human liver buds into mice. To suppress immune rejection, these mice had been genetically engineered to have a dysfunctional immune system. In the initial experiments, the researchers transplanted the buds to the cranium of the mice, which sounds perhaps like a weird and even grotesque choice but was made purely because it was then easier to examine the subsequent growth. That choice of transplant location shows, however, that bodies can be remarkably tolerant of what goes where, once the developmental process is well underway. For indeed, the liver buds not only survived but began to grow blood vessels in response to the signals received from the mouse tissues. These were *human* blood vessels, made from the endothelial cells already in the organoids. The grafts also took and grew when implanted in the intestinal-abdominal region of the mice (the so-called mesentery), round about where a liver really should go.

Of course, a mouse-sized organ can only do a mouse-sized job. But Takebe and his colleagues have since scaled up their process to make large batches of the organoids: what they call "a manufacturing platform for multicellular organoid supply".

Once implanted with a human organoid, the mice in these experiments are not hybrids (like mules) but *chimeras*: organisms containing the cells of more than one genotype. In a mule, all the cells have the same genome, composed of a mix of horse and donkey genes. But chimeras are genetic mosaics.

We saw earlier (page 82) that some humans are chimeric: their bodies are mosaics of cells with different genomes. For example, some of their cells may come from the mother, having found their way into the fetus through the placenta. But just as the mythical beast of that name exemplified, chimeras can contain tissues of unrelated *species*. The ancient Greek Chimera was described by Homer in *The Iliad* as "lion-fronted and snake behind, a goat in the middle".

The viability of chimeras (in the biological, if not the Homeric, sense) seems less puzzling when we acknowledge our cellular nature. Sex is a very particular and rather specialized way of cells proliferating, and it mixes the genomes of the parent cells. Now, there is no prohibition on a gene from one species being incorporated into the genome of another; industrial genetic engineering of bacteria relies on that. But intimately blending the entire genomes of two species is in general too much for biology to handle: the resulting set of "instructions" makes no sense. That's why egg cells have a mechanism for checking that an arriving sperm is of the right species before the two can merge. You can't make a chimera like the Minotaur by the mating of a woman and a bull.

If there is no actual merging of genomes, however – if every cell in an organism retains its own genotype – then there's no problem. If the individual cells are viable, all that then matters is whether they can get along together. And we know cells of very different types can do that: it's why bacteria thrive in our gut, and indeed all over our bodies. Chimeras are simply another kind of diverse yet harmonious cell community.

* * *

Growing tiny human organoids like Takebe's proto-livers in mice is a remarkable feat, but it's not clear how you could ever grow a complete, full-sized human liver, gut or brain that way.

But what if they were grown in human-sized animals – in a pig, cow or sheep? The idea seems feasible in principle. In practice, needless to say, it's complicated.

For one thing, the whole enterprise sounds ethically challenged. Should we rear animals like this as mere carriers of human spare parts? The yuck factor is considerable: many people will find the notion of a pig with a human liver to be disturbing, even obscene.

Is it possible in the first place, though? The scenario would run something like this: you make induced pluripotent cells from a patient in need of a new liver, kidney, pancreas or whatever, transplant them into a pig embryo, and hope they grow into the corresponding organ in the piglet, to be harvested when – forgive the harshness, but such is the reality of livestock farming – it is time to make bacon. But why would the human stem cells choose the particular fate you want, and not some other?

Several years ago, Japanese biologists Toshihiro Kobayashi and Hiromitsu Nakauchi saw a way to guide that decision. The idea is to create a "niche" for the required organ in the host animal: engineering the embryo so that it lacks the ability to make the organ itself. The idea stemmed from work in the early 1990s in which mouse embryos lacking a gene that was crucial for the development of white blood cells involved in the immune system were injected with embryonic stem cells taken from other mice that had this gene. The embryos developed into mice capable of producing the white blood cells using the injected cells.

This shows a remarkable smartness on the part of the host embryo. The donor stem cells aren't initially committed to a particular fate: they could develop into any tissue type. But as the embryo develops, it's as if its cells say, "Time to make our immune cells – but wait, we don't have the right gene. Ah, but look, here are some foreign cells that do. Let's give the job to them!"

That is, of course, quite a burden of anthropomorphization for cells to bear. Sometimes, though, it is hard to tell a comprehensible story without a narrative like this, precisely because cells display the kind of responsiveness that looks like foresight, intelligence, collaboration. It is this "looks like" that holds the tension about how we think of identity and autonomy in life.

The point is that growing organisms will indeed display the initiative to fill an empty "tissue niche" with competent cells taken from another organism. In 2010 Kobayashi, working at the Gurdon Institute in Cambridge, showed that the process could be engineered: a niche can be made to order. He and his colleagues knocked out a gene crucial for pancreas development in mouse blastocysts, and then added to these embryos some cells taken from another mouse in which that gene was still active. Sure enough, the embryos developed into mice with a pancreas. Two years later, Kobayashi and Nakauchi showed that the same trick worked for making kidneys.

Making it work across species looks like a challenge of another order. The researchers first "broke the species boundary" by growing pancreases from rat embryonic stem cells in mice that lacked the pancreas-making gene, and vice versa. In other words, the resulting chimeric rodents had pancreases made entirely of cells from the other species.*

Experiments on these mouse–rat chimeras have revealed something that I find extraordinary. When rat pluripotent cells were added to embryonic mice, they were found to form part of the resulting mice's gall bladders. *But rats don't have gall bladders.* How can rat cells make an organ that rats themselves don't have?

---

* Note that the mice with pancreases made earlier from embryonic stem cells taken from a different mouse embryo were also chimeras, but not *trans-species* chimeras. Such chimeric mice were first made in the 1960s by IVF pioneer Robert Edwards and others, by allowing embryos to fuse together before the blastocyst stage.

Apparently, these cells have an incipient ability to make gall-bladder tissue, and this ability gets unlocked by the "time to make a gall bladder" signals coming from their environment within the embryonic mice. That ability didn't come from nowhere: the common evolutionary ancestors of rats and mice did have gall bladders, but rats lost them. Their cells, however, "remember", as if preserving the memory of their evolutionary history. Might our own cells, then, hold a potential to make body parts that are *not human* – developmental memories of our own evolutionary ancestors?

So organ growth via trans-species chimerism works for rodents. What about humans and pigs? This is tougher for several reasons. One is that the experiments take longer: pigs have a three-month gestation period, as opposed to about three weeks for mice. And humans and pigs have been on separate evolutionary trajectories for much longer than mice and rats.

The first step was to show that organ niches can be engineered in pigs. In 2013, Kobayashi and Nakauchi reported that pig embryos lacking a gene vital for pancreas development could indeed generate a pancreas from the embryonic stem cells of another pig. The male pigs grown this way could even be used as a source of sperm to make more "non-pancreas-making" pigs by IVF, with a ready-made pancreatic niche.

Could this niche be filled by human stem cells?

It took a four-year project to find out. In 2017, biologist Juan Carlos Izpisúa Belmonte of the Salk Institute in California and his co-workers reported that they had made pig fetuses that contain human cells. The human iPSCs were introduced to pig embryos at the blastocyst stage, and the embryos were allowed to develop for up to four weeks.* (The researchers allowed them to go no further to avoid an ethical outcry about mature pigs with human tissues.)

---

* This general approach of growing human stem cells within animal embryos

Although the survival rate of the human cells was rather low, some were still present, and were developing into muscle and the progenitors of other organs. The team also found that human iPSCs could survive in cattle embryos, albeit again rather inefficiently.

This remains some way from showing that human organs will grow in pigs. But given the experience so far, I believe it can happen – in principle. Should it?

* * *

Izpisúa Belmonte and colleagues were unable to use US federal funding for their work, because in 2015 the National Institutes of Health placed a moratorium on supporting such research until the ethical questions had been carefully considered. Although the NIH had promised to review this decision in 2016 after consultations, the prohibition remains in place at the time of writing – although, given the current incumbent of the White House, nothing is predictable.

No one was more dismayed at the NIH's decision than Nakauchi, who had already moved to Stanford University in California from the University of Tokyo in 2014 to escape a Japanese ban on research that involved the insertion of human pluripotent stem cells into non-human embryos. These prohibitions are deeply frustrating to researchers like Nakauchi who are sure that the technique can be made to work. "Animal-grown organs could transform the lives of thousands of people facing organ failure," he has said. "I just don't understand why there continues to be resistance."

---

with a prepared niche had been demonstrated previously. In 2016, for example, Rudolf Jaenisch and his co-workers made chimeric albino mice from embryos into which they added human "neural crest" cells derived from either iPSCs or embryonic stem cells. Neural crest cells are progenitors of a range of different mature cells, among which are melanocytes: pigment-producing cells, which cause hair colouration. The human cells survived in the developing mice, which, once born, had black patches of hair courtesy of their "humanized" tissues.

But that resistance is surely not so hard to fathom. Chimeras seem to upset the natural order. That was their mythical role, after all: the Chimera of Greek legend manifests as a fire-breathing omen of ill fortune. It is a "monster" in the etymological sense: a signifier of a rift in nature. Biological deformities in the Middle Ages and the Enlightenment, including misshapen human babies, perpetuated this role as harbinger: they were not mere aberrations, but portents. They warn us to beware.

Chimeric organisms in the modern, technical sense aren't just reminiscent of these mythical origins; they are representations of them. They do actually blend tissues across species. After the NIH imposed its moratorium on this research in 2015, it conducted a public consultation in which a majority among the several thousand people who expressed a view were opposed to research on chimeras. And these misgivings are not wrong; they are not ill-informed, knee-jerk, Luddite reactions. (At least not entirely; many people had the mistaken impression that the creation of chimeras would necessarily involve *human* embryos.) There is only so much re-arrangement of people's intuitions about the "natural order" that they can be expected to accept all at once.

This is not to say that we must resign ourselves to such judgements. Personally, I do not believe that a ban on chimeras for regenerative medicine is the right decision, but I can't adduce any philosophical calculus in support of that view. I do feel uneasy about the notion of pigs being grown as vectors for human organs, to be slaughtered and dismembered when their time has come. But I realize that this is an illogical position for someone who eats pork and bacon, where the slaughter serves no purpose beyond the gratification of our gastronomical cravings.

I don't, however, fear that there is something unnatural, some violation of propriety, in the sheer existence of a pig with a human liver. Such a combination runs deeply against our experience, and it can easily morph into much more disturbing visions. Yet our instinctive reaction is one based on an illusion, namely that the human is an integral and inviolable whole, a well-defined entity with a homogeneous biological identity. Once we come to recognize that we are co-evolved, co-developed communities of cells, a pig–human chimera seems no more unnatural and repugnant than the notion that our gut is colonized by symbiotic bacteria. The question is simply: do the cells get along?

The real issue, it seems to me, is less the *what* than the *how* and the *why*. This is not a matter of means justifying ends. Quite the opposite – to navigate the astonishing and sometimes frightening possibilities that cell biotechnology is creating, we should be wary of absolute pronouncements about what is right and wrong, what is natural and unnatural. We should ask instead how well we will be served, as individuals and societies, by what we choose to do. Ignoring animal welfare in order to save lives will in the end be morally corrosive, but so will a refusal to alleviate human suffering on the tenuous grounds that it "feels wrong" to some groups.

This is the problem with being a community of cells that has evolved a sense of unique identity and moral agency. It's not easy to fit those two characters together. We might do so most wisely and humanely by denying neither of them.

* * *

Even researchers who are impatient to see this technology forge ahead recognize that there are limits to what might be acceptable, at least without very careful ethical consideration. Kobayashi and Nakauchi have suggested what some of those limits might be, and their list is eye-opening. They say we should think twice before launching into any of the following:

1. Extensive modification of the brain of an animal, by implantation of human-derived cells, which might result in altered cognitive capacity approaching human "consciousness" or "sentience" or "human-like" behavioural capabilities.

2. Situations wherein functional human gametes (eggs or sperm) might develop from precursor cell types in an animal, and where fertilization between either human (or human-derived) gametes and animal gametes might then occur.

3. Cellular or genetic modifications that could result in animals with aspects of a human-like appearance (skin type, limb, or facial structure) or characteristics, such as speech.

We know these three scenarios already. They are the images, the dreams and nightmares, of legends and fiction: (1) takes us to the Island of Doctor Moreau; (2) shows us King Minos's wife Pasiphaë, under Poseidon's enchantment, mating with a bull to spawn the Minotaur; and (3) is the Centaur.

Kobayashi and Nakauchi present this list of possibilities in a paper called "Revisiting the Flight of Icarus", for they suggest that by designing and fashioning for himself wings "to achieve his ambitious goal of flying", Icarus was "chimerizing" his own body by adding to it the desired part of another species. I guess they did not really know the myth – for of course the wings were made by Icarus's father Daedalus, and not simply to be able to fly but in order to escape imprisonment by Minos on Crete. Daedalus was incarcerated by the furious king because he had made the artificial cow-structure within which the enchanted Pasiphaë hid herself to have union with the bull. Researchers in this field are likely to find themselves increasingly grappling with myth and dealing with what myth represents. So it might be wise for them to include among their reading lists not just *Cell*, *Nature* and *Science*, but Homer and Robert Graves.

Now, let me be clear that Kobayashi and Nakauchi immediately followed up their list by saying that they are confident none of these outcomes will ever be realized. There seem to be barriers, they say, that will limit such large contribution of cells, tissues and body parts from the donor in trans-species "xenotransplanation" experiments. Besides, there are various measures one can take to ensure that inadvertent colonization of the host body by the donor tissues doesn't happen.

Perhaps. But not everyone feels that the limits of the possible (let alone the permissible) are so clear. Take a 2013 study in which the progenitor cells of human brain (glial) cells were grafted into newborn mice. When they had matured, the mice showed better learning and memory, for example in how to navigate mazes. This doesn't mean that the mice had acquired more human-like cognitive powers – but the researchers suggested that the greater complexity and capabilities of the human glia stimulated neuronal processing in the mouse brain networks. In truth we don't really know what has gone on in the mouse brains to produce these improvements in performance, but nevertheless it does seem fair to say that the presence of the human brain cells made the mice smarter.

I have seen experts on brain organoids talk with all seriousness about the prospect, although certainly not the desirability, of a brain made from human neurons grown inside a pig: a full-sized brain, blessed with a vascular system like the liver buds that Takanori Takebe has grown in mice. Again, consider it merely a thought experiment. What, then, should *we* think of it? What would a humanized pig think?

I'm not talking about a research proposal, and if anyone were crazy enough to suggest it, it would be rightly rejected. I'm talking about how biological advances are dismantling old certainties and contriving new possibilities – and prompting challenging questions about where we draw lines, and why, and how.

CHAPTER 6

FLESH OF MY FLESH

QUESTIONING THE FUTURE
OF SEX AND REPRODUCTION

As I watched the progress of my mini-brain through the microscope, it was not the first time I'd seen cells with my genes, alive in a glass dish.

But the ones I'd seen previously weren't exactly mine. Only 50 per cent of their genes had anything to do with me. The rest came from my partner. They were IVF embryos, divided at around the four-cell stage.

Did I anthropomorphize those embryos, imbuing them with personality, casting them as plucky little characters determined to give it their best shot at becoming a baby? You bet I did.

What *were* these entities, though? Potential people, some might say, over which I have no right to assert any sort of ownership (except legally). Yet it turned out that they were *not* potential people. For whatever reason, even those that had looked most promising didn't have what it takes to become a baby. Perhaps they just didn't get lucky, but it's more likely that in some way or another the biology was flawed. As we saw earlier, most human embryos at this stage – just two days after conception – prove to be unviable, whether they are conceived *in vitro* or in the body. So it's not at all clear how much scope there is for improving the current success rates for IVF, which stand at typically around 20 to 30 per cent for healthy women under 35.

In vitro fertilization changed the way we can procreate, but it did more than that. It turned our view of ourselves upside down in a way that is rarely acknowledged. It showed us our cellular beginnings, complicating the distinction between the cell and the person. It separated sex from procreation, and it did so at a time – in the late 1960s and the 1970s – when sex, gender and family roles were being subverted and reconfigured as never before. IVF made sex optional, just as the contraceptive pill made procreation optional.* It transformed our ability to study the human embryo. And it produced a host of questions on which traditional ideas about society, marriage, morality, biology and reproduction offered little guidance. We are still finding our way through this new landscape of techniques for growing humans.

Embryology has revealed the embryo to be an ambiguous and protean thing. In one view, it is a colony of cells "exploiting" the maternal environment to its own benefit. Especially in its earliest stages – as a blastocyst, say – it seems more akin to "human tissue" than to a human. It may exchange cells with the mother. It is certainly not, at any rate, an autonomous, independently viable organism. Its future is contingent, fragile, uncertain, and quite possibly doomed.

All the fractious and furious arguments about the "moral status of the embryo" arise from the incommensurability of cell and person. When, thanks to IVF technology, we became able to see

---

* This was no coincidence. The understanding of the role of hormones in the female ovulation cycle that emerged from research on assisted conception and IVF informed work on artificial ways to control that cycle for contraceptive purposes. Some scientists were engaged in both fields at the same time.

and even intervene in the earliest stages of an individual's life, we discovered that we no longer had adequate notions of personhood needed to think about the status of that living entity. Some of us may now see images of their early "selves" as clusters of fewer cells than can be counted on the fingers of two hands; as embryos produced in a dish, ready to be introduced to the womb.

So the anguished ethical debates that surround assisted conception are much more than attempts to establish the ground rules of appropriate legislation. They are efforts to grapple with this re-imagining of what it means to be a human being. That cluster of cells was once all we were. At what point did they become us?

* * *

We've known for at least several centuries that the physical act of sexual intercourse isn't essential for reproduction. The first recorded instance of human artificial insemination was performed by the Scottish surgeon John Hunter in the 1770s, who allegedly made a woman pregnant this way with her husband's sperm. Rather better documented was the procedure in which the American doctor William Pancoast used donor sperm in 1884 to impregnate a woman while she lay unconscious under general anaesthesia with chloroform.* Having inspected the husband's sperm under a microscope, Pancoast knew that the man was infertile, and he apparently considered that he was doing the couple a favour. Neither of them knew about the operation at the time. The doctor told the husband later; the wife was never informed.†

---

* "Donor sperm" is something of an anachronism here. Accounts say that the sperm came from one of Pancoast's students, who was generally agreed to be the most handsome. The students were sworn to secrecy.

† The reasons for not telling her are unknown. While the ethics of the procedure seem rightly to be horrifying today, that has tended to foreclose any deeper enquiry about the affair. Was it thought that the mother would not love her child

The microscope was at that time disclosing the biological moment of conception. The entry of a sperm into a human egg was first observed in 1879 by the Swiss zoologist Hermann Fol. This apparently did not produce an embryo, but growing embryos was one of the first things researchers attempted after Carrel and Burrows refined the art of tissue culturing. In 1912, American anatomists John McWhorter and Allen Whipple found that they could keep three-day-old chick embryos alive *in vitro* for up to 31 hours. A year later, the Belgian embryologist Albert Brachet showed that he could sustain rabbit embryos at the blastocyst stage in a dish.

Creating a viable embryo from gametes outside the body – genuine *in vitro* fertilization – was another matter. Rabbit embryos made by IVF were reported by the American biologist Gregory Pincus in the 1930s, and he even claimed in the following decade to have done the same with human eggs and sperm – but these findings were never verified. The first convincing report of IVF as a reproductive procedure for mammals appeared in the 1950s from Pincus's collaborator, Chinese-American biologist Min Chueh Chang, who described the birth of live rabbits conceived this way. To verify that they had truly been produced by IVF, Chang colour-coded them: he combined the eggs and sperm of black rabbits and transferred the embryos to a white rabbit for gestation. The baby rabbits were black.

Human fertilization is harder. It isn't just a matter of throwing eggs and sperm together and letting them do their thing. As we saw earlier, it is a complex process that requires some intervention of the female reproductive organs, and for decades no one could find a way to make it happen outside the body. Simply not enough was known about the basic biology of fertilization.

---

if she knew the truth? That she would find the process too shocking and shameful? That she would denounce and condemn the doctor and students? Was this paternalistic chauvinism pure and simple? The truth would surely provide a useful datum in the evolution of public attitudes to assisted conception.

American gynecologist John Rock was determined to obtain a fertilized human egg to examine at the very first stages after conception, and he initiated a programme in the 1930s that today almost beggars belief. With his assistants Arthur Hertig and Miriam Menkin, he searched for fertilized eggs in volunteers scheduled for hysterectomy. The researchers hinted to the women that they might have sexual intercourse just before the operation. That the patients agreed to the procedure testifies to an extraordinary generosity in the cause of furthering an understanding of fertility and its obstacles. That the research was permitted at all shows how little recognition there still was of the need for ethical regulation of medicine.

In 1944, Rock and Menkin claimed to have achieved the first human *in vitro* fertilization, from eggs collected during surgery.* The researchers were able to watch the fertilized eggs begin to divide, but that's as far as they got: they couldn't claim to see a genuine embryo in the petri dish. Rock went on to make pioneering contributions to the development of the oral contraceptive pill.

There was something of a Wild West ethos in the early years of IVF, when speculative research relied on boldness, persuasion and a certain amount of hubris. While working at the National Institute for Medical Research in north London in the 1960s, physiologist Robert Edwards obtained eggs from any surgeons and gynecologists he could find who were sympathetic to his aims. They were acquired in the course of ovarian surgery, and the "donors" were not asked for consent. Such liberties were the order of the day, but although Edwards was motivated by a strong desire to alleviate the anguish of infertility, it's hard not to see in this image of the male doctor experimenting on unsuspecting women to initiate "new life" a

---

* It complicates the story even more to note that Rock was a devout Catholic, albeit one who evidently disagreed with the church's views on birth control and attempts at assisted conception. His search for fertilized eggs in utero could be regarded, after all, as a kind of abortion.

cultural attitude that had not advanced much since Pancoast's time. As anthropologist Lynn Morgan has pointed out, the history of embryology is characterized by the anonymity of the women from whom embryos and eggs have come: they are often treated as a passive, faceless source of biological material for study. The wariness or even antipathy of some feminists towards reproductive technologies may stem from a well-founded concern that they retread narratives of male control and dominance.

Fame and glory did not feature high on Edwards's agenda, however. On the contrary, his efforts courted attacks and ridicule from his peers. His PhD student Martin Johnson summarizes the atmosphere in which his team worked:

> If I'm honest, while we were doing our PhDs, and even into our postdoctoral time in the lab, [we] were very unsure about whether what Bob was doing was appropriate, and we didn't want to get too involved in it. The reasons for that were partly because it was quite unsettling as graduate students and early postdocs to see the sheer level of hostility to the work – when Nobel Laureates and the Fellows of the Royal Society and the emerging bigwigs of the subject . . . were lambasting into [sic] Bob and saying, you shouldn't do it . . . you had to say, well, what's going on here?

Despite the scepticism, antagonism and ethical disapproval of his peers – eminent biologists such as James Watson and Max Perutz later warned that IVF might produce babies with serious birth defects* – and the refusal of the UK's Medical Research Council to fund his work, in 1969 Edwards, working with gynecologist

---

* When you bear in mind that animal IVF experiments had shown no such thing, it becomes more clear that these fears were coming from some source other than their informed scientific judgement.

Patrick Steptoe and Edwards's student Barry Bavister, published in *Nature* a detailed account of human sperm entering an egg *in vitro.*** "Fertilized human eggs," they wrote, "could be useful in treating some forms of infertility." The following year Edwards, Steptoe and their clinical assistant Jean Purdy published images of fertilized human embryos grown to the 16-cell stage; by 1971 they had developed embryos *in vitro* to blastocysts.

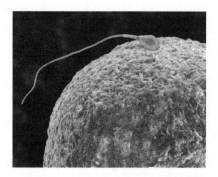

The moment of conception? Sperm on the threshold of entering an egg.

Steptoe had the surgical knowhow to re-implant such embryos in the womb, and the researchers knew there would be no shortage of volunteers for this highly uncertain and perhaps even dangerous procedure.

But these photos of human embryos in a dish had a deeper significance. For the first time, we could we see the start of the journey that had been previously tracked back only to something recognizable as a kind of homunculus, a blob of tissue in which a shrimp-like head was already forming.

---

* The paper was published on Valentine's Day, a gift to press interest. That could hardly be the authors' doing but was precisely the sort of canny decision for which *Nature*'s then editor John Maddox was renowned, and I have my suspicions.

We have been able to think about ourselves in developmental terms only to the extent that we could *see* that process. "Embryos as we know them today are a relatively recent invention," says Lynn Morgan. "A hundred years ago, most Americans probably would not have been able to conjure up a mental image of a human embryo." Morgan points out how some cultures did not regard the tiny entities sometimes revealed by spontaneous abortion as truly human, and as possessing the same moral status as people.* Many "pro-life"† groups now argue their case by mobilizing the imagery of biomedical technology, using the *in utero* fetus as a stand-in for the embryo to imply the presence and continuity of personhood from conception onwards.

According to historian of science Nick Hopwood, the concept of human development has been actively constructed, not simply uncovered as a "fact of life". That process, he says, began in earnest with the embryology of the late nineteenth century. Biologists and physicians presented embryogenesis as a complex yet unremarkable biological process, and it was often assumed that the moral questions that it raised could be clarified and even resolved merely by better scientific understanding.

We now know that this is not the case; if anything, the reverse is true. It was hard to know how to make sense of the ontogeny that IVF revealed. By what somersault of the imagination can we link the bawling baby with a little knob of cells resembling nothing much more than a cluster of soap bubbles?

* * *

---

* Even newborn babies were not always perceived as yet fully human – a fact occasioned by the high rates of infant mortality before the twentieth century, which perhaps even demanded a degree of psychological distancing in the early days of infancy.

† I use that terminology precisely to emphasize the politicized language; the phrase is not descriptive.

We tried to make that connection with a phrase that yoked together the archetypal apparatus of the chemical laboratory – the arena of the artificial – with the sacred product of conception and birth. We began to speak of the "test-tube baby".

Test-tubes were never involved in IVF; their role here is purely symbolic. The phrase "test-tube baby" was first coined in the early twentieth century, when a public understanding of biology was so rudimentary that the *de novo* creation of life itself from mere chemistry sounded entirely feasible, even imminent. In that circumstance, what we today call IVF – initiating conception and perhaps growth too outside the body – was a feat that looked not so far removed from a god-like creation of life itself.

The child housed in a glass receptacle has long been a part of the imaginarium of human life and death. Babies stillborn, miscarried and deformed have been preserved in jars and bottles for centuries. As Susan Merrill Squier has documented, the image of a being not just stored in a bottle after death but actually created in that vitreous, artificial environment goes back at least to the homunculus that alchemists and mystics in the Middle Ages and Renaissance claimed to be able to make in the laboratory. There are recipes for doing it; Goethe's *Faust* describes the genesis of such a creature and illustrates the moral context in which they would be evaluated.*

The *in vitro* culturing of cells transformed that narrative and gave us the test-tube baby. In his 1924 book *Daedalus, or Science & the Future*, J. B. S. Haldane described the promise of ectogenesis (*in vitro* gestation), inspiring his friend Aldous Huxley to pen his famous satire a decade later. In the future society of *Brave New World*, ectogenic babies are chemically manipulated to create a graded caste system of intelligence.

---

* I tell that story in my book *Unnatural*.

Haldane saw this (putative) technology as a blessing. It could support both the emancipation of women – which he welcomed in principle – and eugenic social engineering to preserve the vigour of the human race. Haldane and Julian Huxley worried that the widening opportunities for women were making the more educated and intelligent among them less inclined to bear children as they discovered that there could be more to life than domestic drudgery. Lacking such prospects, the "lower classes" went on breeding regardless, with the result (Haldane feared) that the gene pool was degenerating. As Haldane's narrator in *Daedalus* explains from his future vantage point in the twenty-first century:

> Had it not been for ectogenesis there would be little doubt that civilization would have collapsed within a measurable time owing to the greater fertility* of the less desirable members of the population.

In this way, making people artificially in the controlled environment of the laboratory played to concerns in the inter-war period about population control and the decline of civilization.

Haldane never imagined that everyone would welcome his vision. "There is no great invention, from fire to flying, which has not been hailed as an insult to some god," he wrote. "But if every physical and chemical invention is a blasphemy, every biological invention is a perversion." He knew that some would call ectogenesis and related techniques for manipulating conception in the laboratory "indecent and unnatural", and he was right. In her overwrought 1938 article for *Tit-Bits* inspired by the tissue-culture research at

---

* Haldane indulges the unfortunate tendency to use "fertility" to mean the actual number of offspring, rather than the potential to bear offspring. This confusing usage is pretty much universal now, but with no lessening of its liability to mislead.

Strangeways (see page 186), Norah Burke spoke of "chemical babies" and asked, "What sort of creatures will these be?" The article's title guided the reader towards the desired recoiling response: "Could you *love* a chemical baby?"

But the freighted term "test-tube baby" was itself minted, it seems, by none other than Thomas Strangeways. "It will thus be seen," he said in his 1926 lecture on tissue culture, "that the idea of the 'test-tube baby' is not inherently impossible." It was a more resonant term than Haldane's ectogenesis, which sounded like scientific jargon. Here Strangeways conjured up an image anyone could appreciate, at the same time astonishing, exciting and terrifying. It is an icon of modernity itself: of humanity in the age of the scientific control of life.

A stock representation of the "test-tube baby", showing how the recognizably human infant form stands in for the pre-blastocyst embryo, which is in fact what is produced in the in vitro *stage.*

In short, "test-tube baby" was the right phrase for the right time. That a human could be a product of artifice seemed almost an inevitable conclusion of industrialized mass production, which after all seemed to be generating everything else in people's lives on an assembly line, standardized, tested and commodified.

It is a short step from Haldane's ectogenesis to Aldous Huxley's Central Hatchery, but perhaps even more conforming with Strangeways's phrase is the idea of the *robot* as it was first conceived by Czech writer Karel Čapek in his 1921 play *R.U.R.* The abbreviation is the name of a company, Rossum's Universal Robots. But whereas Čapek's robot (the Czech for "labourer") today conjures up humanoid devices of metal and wires – the Terminator with its synthetic skin stripped away – Rossum's robots weren't like that at all. They were made of pliant flesh.

The director-general of R.U.R., Harry Domin, explains that old Rossum, the inventor of the robot, arrived at his discovery via chemical experiments conducted in test-tubes that aimed to create a living substance. Rossum was a marine biologist* who discovered a new form of "protoplasm" that was considerably simpler in chemical terms than the stuff found inside cells. "Next he had to get this life out of the test-tube," says Domin.

From this artificial living matter, the company fashions a kind of dough that can be shaped into organs. "There are the vats of liver and brain and so on," says Domin. "Then there's the assembly room where all these things are put together." The process owes a debt to Henry Ford's factory automation, but it's obvious that the underlying technology is that of tissue and organ culture pioneered by Carrel and the Strangeways researchers.

The same fears that Čapek mined – of the homogeneous, industrial-scale automation of reproduction – drove David H. Keller's story "A Biological Experiment" in a 1928 issue of *Amazing Stories*. It anticipates *Brave New World* with its dystopian future society in which sex is forbidden and babies are grown in factory vats

---

* This is a reference to Jacques Loeb, a German-American biologist who was alleged to have created life chemically when, in experiments in the United States in the 1890s, he discovered how to induce parthenogenesis in sea urchins by treating them with chemical salts.

according to a standardized specification engineered by radiation treatment, to be dispensed to couples who have been granted the requisite government permit.

Dystopias of course always make for better stories than utopias. All the same, it is significant how often these tales of engineered "chemical babies" end with the conquest by the manufactured beings of the rest of humanity. *R.U.R.* established the template for tales of malevolent robot rebellion and conquest that persist today with *Westworld* and Skynet in the *Terminator* series; fiction has little use for a robot that always does what it is told. The unspoken assumption here is that there is something inherently amoral and ruthless about people made by artificial means. In Norah Burke's exposé, the "chemical babies" are portrayed, for no apparent reason, as "sexless, soulless creatures of chemistry" that may in the end "conquer the true human beings" and lead to an "end to humanity". But perhaps it is not so hard to understand where, in England in 1938, such fears were really coming from. "How to grow a human" is not and has never been a question for science alone but is deeply and inevitably a sociopolitical affair.

The test-tube baby trope wasn't confined to *Amazing Stories*; it was just as apt to appear in the august pages of *Nature*. Its genesis shows why it is quite wrong to suppose (as too many scientists do) that science just gets on with its sober work while media and popular culture come along and traduce it with sensationalist slogans and images. The truth is that the "professional" and "popular" faces of scientific innovation co-evolve. Honor Fell's enthusiasm for popularization and dissemination of the research at Strangeways – motivated in large part by a desire to win support and funding for the research – dissipated as she saw the lurid headlines and science-fiction stories that it generated. Alarmed by rumours that the Strangeways lab was aiming to produce test-tube babies, Fell insisted in 1935 that tissue culture should be presented by scientists only as "a valuable technique with peculiar advantages and limitations". That did not prevent the *Daily Express* from writing the following year that at Strangeways "living tissues are growing and developing exactly as they would in the complete parts of living animal bodies." A quote in that article from a suspiciously anonymous scientist "from another Cambridge lab" claimed that the research takes "the first steps to the *Brave New World* visualized by Aldous Huxley, with babies cultivated in test tubes". As historian Duncan Wilson points out, while Fell was happy to write and broadcast about the work at Strangeways, being cast as the creator of soulless chemical babies was "clearly not the sort of propaganda that she had in mind".

One might be tempted to point the responsibility for those distortions and exaggerations back at Fell. But she was surely right to suggest that scientists *should* talk about what they do, and try to make it relevant to their audience. The point is that they must recognize that they will never keep control of their narratives and metaphors, so they had better think rather carefully about which ones they select. The same tensions are apparent today in discussions of genetics and genomics. The scientists deplore the simplistic genetic determinism that has evolved in the public view of the subject, but it is an easy matter to justify a response that says, "Well, you started it."

* * *

I met Louise Brown – the first person to have been born by IVF – on her fortieth birthday in 2018, and it brought home to me how recently it is that we found a new way of growing people. Louise is a relative youngster; I was preparing to head off to university when she was born. Her eldest son was just 11 years old on her fortieth anniversary.*

---

* His birth didn't involve IVF, negating the common preconception – no more,

Saying that Louise came from a "new way of growing people" might sound derogating, as if she stepped from the pages of Aldous Huxley. But we were all grown. It's just that, before 1977, the process had never begun outside the body. I put it this way to remind us that we were each one of us unfolded piece by piece from a fertilized egg in an orchestrated proliferation of flesh. Louise's conception, and what followed from it, played a big part in making that process explicit and visible, forcing us to grapple with what we could previously wrap in tradition and myth, euphemism and dogma. Science has a way of doing that.

It's easy to forget how much was invested, before IVF, in a contingency that turned out to be merely a matter of finding the right biochemistry. Back then, there was a common feeling that the act of conventional sexual intercourse – within marriage of course! – not only produced but sanctified the child. "[The] tradition," wrote *Life* magazine's science editor Albert Rosenfeld in its 1969 issue on "science and sex", "has been to regard a child as the product of the marriage bed – and therefore, in some way, sacred." It was as if the coy phrase "married love" were some vital ingredient surrounding the gametes like an aura and ensuring the normalcy and propriety of the child. But, Rosenfeld warned:

> the force of love may henceforth have little to do with the process. The crucial fragments of the world may simply be taken out of cold storage on demand . . . love in the old sense would no longer be a part of the procreative process.

The sentiment here is much the same as that which led Pope Pius XII to condemn artificial insemination in the 1940s on the grounds

---

in fact, than an ancient, quasi-magical prejudice – that a person conceived "artificially" would be infertile.

that it would "reduce . . . the act of married love to a mere organic activity for transmitting semen", turning "the sanctuary of the family into a biological laboratory". Rosenfeld's "force of love" here indeed becomes almost an active biological agent, directing the outcome of cellular union like a spiritual transcription factor. Barely 50 per cent of the Americans polled by *Life* in 1969 imagined they "could feel love towards a baby" conceived by IVF from their gametes; only a few per cent more imagined that a child so conceived would love its parents. In the test-tube, it seemed, love evaporated.

All this created a sense of cognitive dissonance around Louise's birth in 1978 – an event so familiar, so ordinary, that reporters did not quite know how to merge it with their science-fictional fantasies. Witness how *Newsweek* announced her arrival into the world by caesarian section at Oldham General Hospital in northern England: "She was born at around 11.47 pm with a lusty yell, and it was a cry round the brave new world." Or this from the *Daily Express*: "She's beautiful – that's the test-tube baby." Pretty much uniquely among people conceived by IVF, Louise is *still* "the test-tube baby". She seems to have resigned herself to it with grace. Meeting her in person shows at once how absurd and misguided all those preconceptions were. The ordinariness of her family life in Bristol – Louise is, I believe, happy and even eager that it be seen as such – contrasts starkly with the strange kind of celebrity status that her then-unique mode of conception "in a test-tube" has given her.

The lexicon of IVF warrants further interrogation. For the newspapers insisted that Louise Brown was something else too: a "miracle baby". And aren't miracles supernatural events?

Yes indeed, and we know who the original miracle baby was, born of a virgin in Bethlehem two millennia ago.

To be enabled to have a child when you didn't think it possible, like Louise's mother Lesley Brown (who had blocked fallopian tubes), certainly seems marvellous and wonderful. But the insistence with which IVF children are even now denoted "miracle babies" speaks of other factors at play. It implies that this is no mere medical intervention in a biological process but has instantiated God's work: an achievement both astonishing and hubristic. Faustian, you might say.

Anthropologist Sarah Franklin sees a quasi-religious connotation to the way IVF is portrayed as a "technology of hope". Hoping for a child was once deferred to the grace of God, who alone decided whether or not to bring it into existence. In the absence of that hope and grace, what you have are souls in anguish: the "desperate" couples so essential to the media narratives of assisted reproductive technologies. That desperation, and its dependence on divine will, have the imprimatur of holy scripture:

> Now when Rachel saw that she bore Jacob no children, Rachel envied her sister, and said to Jacob, "Give me children, or else I die!" And Jacob's anger was aroused against Rachel, and he said, "Am I in the place of God, who has withheld from you the fruit of the womb?"

Infertility was once regarded as a sign of divine displeasure, and it still carries a moral stigma – in secular culture, often transmuted to the idea that women who cannot get pregnant are too uptight, while those who will not are selfish. As biologist Clara Pinto-Correia has said, "There is still not one social group, not one culture anywhere on earth, that doesn't abhor infertility."

But sexual intercourse and procreation was always a complex transaction in the Christian West. What mattered so dearly, so perversely, to the miraculous birth of Christ was that no sin was involved. Which is to say, no sex.* Likewise, so much of the suspicion, prurience and religious condemnation that has surrounded IVF comes from the fact that it can produce a child without copulation. That creates a contradictory mélange of responses: is this a kind of special purity, or is it unnatural? Such uncertainty accounted for much theological unease about the medieval homunculus: not born of Adam, the being might be free of original sin. Before Louise Brown, the most famous person to have been made without sex in the usual sense was Frankenstein's monster, who said to his maker, "Remember that I am thy creature; I ought to be thy Adam, but I am rather the fallen angel."

Deep anxiety about procreation has always led us to corral and constrain sex itself (and has then tempted us to test and flirt with those boundaries). That, I suspect, is the fate of a creature facing Darwinian selection with a brain far larger than is strictly needed for the job: sex will always be more of a problem for us than for our animal cousins. But IVF brought new problems to weigh on the mind and the conscience. Among the many unsettling notions it invoked was that *sex is now optional.* Was that bad? Was it bad because sex is *meant* to be bad, except when (even when?) marriage made it obligatory? It was all very confusing.

* * *

The invention of *in vitro* fertilization was the pivotal moment in changing attitudes about the nature and uses of human cells. By creating more fertilized eggs than are generally re-implanted in the uterus, IVF produces "spare embryos" that may be used, with consent, for scientific research. It has made possible the entire field

---

* Jesus's was not the Immaculate Conception, however – that was the conception of the Virgin Mary, which was occasioned by sex between her parents but granted special, pre-emptive exemption from original sin.

of human embryo research, while the availability of human embryonic stem cells in these embryos has opened up a new arena of
biomedical study. At the same time, IVF permits intervention in
human growth at its earliest stages, motivating new possibilities for
overcoming disease but also fears of how biomedical technologies
might be used to alter our natures.

On the one hand, somewhere currently in the region of 6 to 8
million IVF births* have normalized the technique to the point
where there is no longer any stigma or sense of unnaturalness
attached to it. The Catholic church still officially bans it, but – as
its authorities surely know – many Catholics take no notice. In
some countries, IVF now accounts for around 6 per cent of the
annual births.

On the other hand, I don't believe we have yet culturally processed
the shift that IVF has occasioned in our ideas about how humans
come into being. Rather, we have constructed normative stories
just as energetically as we made the idea of IVF alien and strange
before it became a regular part of the reproductive landscape. Yet
these narratives don't eliminate our prejudices and preconceptions
about how humans are (and should be) made. Those judgements
simply seek a new object, which is found every time some new
technology arrives to push back further the technological boundaries of conception. In place of the "test-tube baby" we now have
the "designer baby", the "three-parent baby", the "saviour sibling"
and more.

It's a common mistake to imagine new technologies as being
solutions to easily stated problems. If we were to assert that the
automobile arose from a desire to travel faster, or the computer

---

* The global figure is hard to pin down. But it will be larger still if we include
all those people whose parents were conceived by IVF, and who would not
otherwise have come into existence. Louise Brown's sister Natalie, who was also
conceived by IVF, became in 1999 the first such person to have their own child.

from a desire to perform complex calculations quickly, hindsight reveals the inadequacy of such formulations. Not only do technologies acquire a life of their own and take off in unanticipated directions, but they – especially the most transformative of them – come to represent something more than themselves, and far more than their inventors ever envisaged. They create new possibilities, but the possibilities are never neutral, ethically, morally, socially or politically. They are shaped and selected by hopes, dreams, fears, by cultural decisions and judgements. Technologies enter the language ("change gear", "foot on the gas", "I'm online", "email me"), they provide alluring new metaphors (neither the genome nor the brain is like a computer, but you'd be forgiven for thinking otherwise from popular discourse), they reveal themselves in icon-laden visual shorthand.

That all this applies to IVF is no surprise, speaking as it did to such culturally resonant issues as procreation and infertility. In retrospect, there was not the slightest prospect that IVF would remain a medical procedure for addressing cases of infertility between married couples, which is what Robert Edwards and Patrick Steptoe intended. Not least, once it became a private and lucrative industry, other forces came into play besides the laudable goal of reducing the pain of infertility within marriage.

Should we try to take this transformative potential into account when we develop new technologies? It's not obvious how that might be done. Even the most perspicacious of inventors and discoverers can't be expected to foresee the psychic and cultural currents that their innovations will tap into. No one knew, before the advent of mobile phones, how much we craved the cocoon of a private mental space in public places, nor how thoroughly we could imagine ourselves insulated within it. We hadn't guessed that we were so anxious to escape our surroundings into a place where we could curate our lives.

But we can learn from experience, and it would surely be no bad thing if the acknowledgement that technologies are much more than solutions to problems was recognized beyond, say, the small band of folks who, typically under the academicized banner of "science and technology studies", make these considerations their business.

In the case of IVF, there should have been ample warning that a "cure" for infertility would activate powerful social forces – for there has never been a time in recorded history when we have not fantasized about the possibility of "making life", and especially making people. Contentious issues ranging from abortion to genetic engineering, feminism to transhumanism are all channelled through that *ex vivo* union of sperm and egg. All are conditioned in one way or another by ideas about our own "essence".

Philosopher Rom Harré caught the crux of the matter when he proposed that "our social identities, the kinds of persons we take ourselves and others to be, are closely bound up with the kinds of bodies we believe we have." Because IVF has changed our beliefs about our bodies, it has changed our concepts of identity.

"Conception in vitro is now a normal fact of life," writes Sarah Franklin, "yet having passed through the looking glass of IVF, neither human reproduction nor reproductive biology look quite the same." IVF, she says,

> confirms the viability of a new technological ground state, or norm, of human existence and renewal . . . The meaning of being after IVF . . . describes an ambivalent position that is constantly being reworked, rewritten, and recomposed.

Franklin argues that unease about the "artificiality" of the moment of *in vitro* fertilization and conception has in fact motivated social conformity in the presentation and practice of IVF. While her assertion that it "is designed to precisely replicate each of the steps in the journey to parenthood" seems a little hard to swallow for anyone who has actually gone through it, the imagery typical of IVF clinics has an assertively heteronormal, fairy-tale quality. And that is not simply a veneer that society and the marketplace has chosen to lay over a "neutral" science. Patrick Steptoe's opposition to IVF for single or lesbian women on the grounds that it was unnatural and morally wrong obviously reflects the mores of his time (from which scientists have never in general been exempt); yet still this reminds us that the architects of new technologies do not work in a sociopolitical vacuum.

As Franklin points out, IVF clinics are keen to reassure clients that the technology is simply giving nature a "helping hand". Sure, there is a bit of stuff that happens in the lab – but soon enough the resulting embryo is back in the uterus where it belongs, and everything goes ahead as normal, positively disguising the more radical implications. You'll rarely see a media article about assisted conception without the obligatory image of an egg cell or few-cell embryo under the microscope. Chances are, it will show the IVF variant called intracytoplasmic sperm injection (ICSI), in which a lack of good sperm mobility – a common cause of fertility problems – is addressed by injecting a sperm directly into the egg using a tiny needle. The message here is complex and coded. With its luminous backlighting and the presence of technical equipment – the needle, and the larger pipette on which the egg cell is held – we are left in no doubt that this is a technological procedure. But the needle penetrates the soft, yielding egg in a cell-scale recapitulation of intercourse, whispering to us the reassurance that this is after all just a familiar sexual union, with the glassware there just to give nature that "helping hand". The message is: Sure, this is a technology – but it looks a lot like real sex, right? In such ways, a wholly technique-driven procedure for making humans is naturalized and can be smoothly accommodated into our culturally approved methods of procreation. The "love force" triumphs after all.

An assertion of traditional gender roles is another of the subtexts of these constructed narratives – and of course, everyone in the adverts is better looking (and probably younger) than the average consumer. One can, I suppose, hardly blame IVF clinics for displaying a parade of gorgeous babies – those after all are the product. But the stock advertising tropes are revealing in themselves: of course these business ventures will emphasize their "superiority" (here measured by the pregnancy success rate) against rival brands. Any company in the parenthood sector does the same. What is odd about this market, however, is that there can be no guarantee that you will get the product you're paying for.

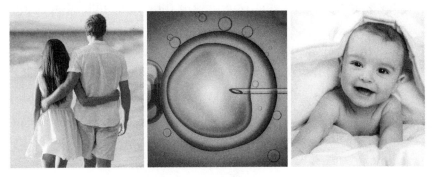

*Love–sex–baby: the reassuringly "normal" visual
narrative of procreation offered by IVF clinics.*

It's often pointed out that the IVF narrative sidelines those for whom it doesn't work, as well as the arduous nature of the process. It remains difficult, uncomfortable and uncertain. The success rate drops rather rapidly with the woman's age. Egg collection for producing embryos is painfully invasive and doesn't guarantee their quality.

The official narrative, not only in the media but also in much academic discourse, also makes little or no mention of the role of the male (whether or not he is to be a care-giving parent), reinforcing a perception that this is a female affair – albeit one originally developed largely by men* and often conducted on women by male doctors and technicians. By tacit collusion, a picture emerges in which not just fertility but also infertility is solely the woman's responsibility.

Should women welcome assisted reproductive technologies (ARTs)? Influential feminist writers such as Donna Haraway and Shulamith Firestone think so. Firestone, author of the groundbreaking feminist text *The Dialectic of Sex* (1970), agreed with J. B. S. Haldane that these techniques could be an emancipating force, albeit only if accompanied by a radical re-definition of gender, sex, parenthood and family. ARTs have the potential to dislodge stereotypes; feminist writer José van Dyck is right on the money when she says that one reason they face such opposition

> is not that they are defying nature, but that they are defying culture. They profoundly challenge our preconceptions of a society that seems "naturally" divided along axes of gender, race, class, age and physical or psychological condition. They also challenge traditional social structures, such as the nuclear family, identifiable races or ethnic groups.

As Franklin has said, "There is a revolutionary purpose to be achieved through tools, machines, instruments, and biology, and indeed through their union." That, of course, is precisely what worries social and religious conservatives about ARTs.

But while the possibility of freezing eggs or embryos is mooted today to delay childbearing so that young women can first build

---

* The role of Jean Purdy, in particular in the early work with Edwards and Steptoe, has been afforded too little attention. She was a central member of the team.

successful careers, one might question whether women should have to accept the status quo that imposes that difficult decision in the first place. Some feminist groups have greeted ARTs with scepticism, dismay and even outrage. They have been portrayed as a male conspiracy, orchestrated by "techodocs", to oppress women or even to eliminate them from the procreative act.* In the 1980s, the German militant feminist group Rote Zora bombed IVF clinics. Without condoning that kind of extremism, I think feminists are right to raise concerns: ARTs could challenge stereotypes, but they also have the potential to reinforce them.

ARTs and the science and technology that have stemmed from them – stem cells, genome editing, and advanced tissue culture – challenge deeply held preconceptions not just about procreation, sex and gender but about identity, self, life and death. It's right that we should be unsettled by them, but this does not mean we should fear them. We can begin by bringing these challenges into the open, where we can talk about them honestly and frankly. The value of looking ahead to where such technologies might lead is not to predict, pre-empt and potentially forestall them, but to interrogate their implications for our ideas, values and morals here and now. To put it another way, Franklin argues that IVF is, for anthropologists (and, I think, for society generally), "good to think with".

* * *

On the fortieth anniversary of Louise Brown's birth, much of the media discussion was about those 6 million plus babies who came after. But as Franklin has said,

---

* Conservative critics of putative ARTs in the 1920s seemed more alarmed by the prospect that they would make men redundant. In the face of new ways of making people, we are all insecure.

Even the somewhat surprising scale of human IVF's expansion worldwide over the past thirty-five years may pale in the wake of its future significance – which will not only be measured by IVF's expansion into genetic disease prevention but must also take into account a watershed point in the very meaning of the adjective "biological" as it becomes increasingly synonymous with technology.

Synonymous is perhaps going too far, but what Franklin is driving at here is the fact that biology is no longer distinct from technology. As we saw, IVF embryos provided access for the first time to a supply of human embryonic stem cells. These in turn led to the discovery of our cells' unguessed plasticity.

Such studies on human embryos and their cells might help to improve IVF itself, as well as revealing new understanding of the process of "natural" conception and development: for example, illuminating the causes of early miscarriage and growth defects. "Research on human embryos has changed our fundamental understanding of the genetics of cell biology," says Alison Murdoch, professor of reproductive medicine at Newcastle University. "As well as helping us understand why fertility of the human species is so uniquely bad, the research has extended applications of human embryology beyond the clinical treatment of infertility into prevention of other medical problems."

Yet as we've seen already, research on human embryos is controversial. The idea of experimenting on "spare embryos" that might have the potential to become a human being (though many do not) is morally repugnant to some people. Others accept that excess embryos are inevitable if IVF is ever going to achieve good success rates, and feel it is better to use them to benefit science and medicine rather than just discarding them.

Where you stand in these issues depends in large part on what ethical status you feel the human embryo should have. That is of course a question with wider ramifications, especially for regulation on abortion. Does the embryo deserve the full rights and protections of a human person? Or is it – at least before reaching the fetal stage in which the central nervous system begins to develop – just a ball of cells, more tissue than person? The scientific picture makes the process of human development from zygote to newborn baby a continuous one, with no stage to which we can uniquely point as the threshold of personhood. Even fertilization itself is not a well-defined event – it takes several hours after a sperm enters an egg for the chromosomes to become united into the genome of the individual. Even then the zygote may yet divide into identical twins. Science can inform the ethical debate, but it can't offer a definitive resolution.

Once IVF became a reality, the British government recognized that regulation was needed to navigate and police these problematic boundaries. In 1982, it convened a committee chaired by moral philosopher Mary Warnock to draw up recommendations. The Warnock committee fudged the matter of the embryo's ethical status, calling it a "potential human being" without explaining if or how this "potential" made it less or equally a moral object compared with a human being delivered into the bright light of the world. The report argued that the human embryo should be accorded "special respect" and not be treated as mere material for experimentation. But what did that mean? That experiments be conducted with a suitably sober countenance? The Warnock report resolved none of these issues – and did not pretend otherwise. Any answers, it said, would be "complex amalgams of factual and moral judgements" – and should be set to one side in favour of a more practical resolution about "how it is right to treat the embryo".

That avoidance of definitive judgements might sound like a failing, but it was rather an inevitability that was put to good use.

The report recommended that research on human embryos be limited only to embryos less than 14 days old. The argument was that at this stage the human embryo develops the feature called the "primitive streak", the first intimations of a spinal column (see page 69). Once it has a primitive streak, an embryo can no longer split into identical twins. So in some crude fashion, this point can be considered the threshold of unique personhood.

But it isn't really. For one thing, a primitive streak doesn't always appear on a human embryo precisely at 14 days. And it is just one of a series of developmental landmarks towards the infant. Certainly, it doesn't signify that the embryo can thereafter experience sensations such as pain. The question of why a "potential human" can be material for experimentation before 14 days but not thereafter was unresolved.

Yet the 14-day rule, which became law in the UK in 1990, established a clear and pragmatic boundary for legally enforceable regulation that allowed embryo research to proceed. Warnock had no illusions about the arbitrary nature of her ruling; the point was that it enabled a framework for research, so that researchers knew where they stood. And besides, at that time – and thenceforth, until very recently – the 14-day rule was in no danger of being breached anyway, because no one could keep a human embryo alive *in vitro* for more than five or six days.

But advances in embryology are now pushing at that boundary, raising questions about whether 14 days is still an appropriate cut-off. In 2016, developmental biologist Magdalena Zernicka-Goetz at the University of Cambridge and her co-workers reported that they had found a way to let human embryos develop all the way up to 13 days *in vitro*. At that point, the researchers had to end the experiment to conform to the law; otherwise, who knew how long the embryo might have persisted?

The trick here was to find some way of mimicking implantation into the uterus, which human embryos undergo around seven days after fertilization. Recall that at this stage the embryo is a blastocyst, consisting of a hollow sphere with the inner cell mass – the stem cells that will form the future fetus – clustered in one part. The outer layer of the blastocyst's cells comprise the trophectoderm, which will make the placenta, while the inner cell mass is made up of the epiblast – the nascent fetal tissue – covered with a layer of "primitive endoderm" that will become the yolk sac. Around day 9, the epiblast develops an inner void called the amniotic cavity. Then the beginnings of the body plan appear during the process of gastrulation around day 14, which is when the primitive streak is formed.

To keep the embryo alive beyond the implantation stage, Zernicka-Goetz's team needed nothing especially fancy beyond the right environment. They found that simply sticking to the surface of the plastic slide holding the embryo is enough to keep it growing, so long as the culture medium is conducive (the researchers used amniotic fluid taken from cows). The entity that results doesn't look exactly like an embryo implanted in the uterus – for one thing, it is flatter. But it shows the formation of a yolk-sac-like cavity surrounded by trophectoderm cells, as well as a structure that looks like an epiblast, complete with its own amniotic cavity. All the normal stuff, then.

These *in vitro*-attached embryos offer a laboratory model for studying the early post-implantation stages of development. But many important events also happen after the gastrulation stage, between day 14 and day 28. Because of the impossibility of probing embryo growth in humans during this crucial time, it's still something of a black box – which might contain valuable insights into human health, disease and malformation. Much of what we know about these stages at present comes only from studies of mice, but there are some important differences between mice and men.

So there's much interest and excitement about the prospect, raised by the results of Zernicka-Goetz and her colleagues, of investigating the post-14-day embryo. But that would be made possible only by changing the existing law.* Warnock has advised against rushing to do so, but others think the scientific benefits could warrant reconsideration. Besides, as we will see later, the Warnock Committee's pragmatic compromise may not be adequate to deal with some of the newest ways of growing a human being, where the conventional trajectory of embryo development might be bypassed altogether, along with its agreed moral milestones. There's no consensus on what to do about it. The question of when cells become people isn't going to go away; on the contrary, it becomes ever more pressing as biological science advances.

* * *

One of the common causes of female infertility is a decline in egg quality with age. A woman who wishes to conceive later in life can currently elect for her eggs to be collected at a younger age and frozen for later use. There's no sign that the procedure carries health risks for a child conceived that way, although in truth it's too early to know if problems might appear when the child reaches adulthood.

But what if you didn't plan ahead, and now in your late thirties you are finding it hard to conceive? What if you have no eggs at all, perhaps because of surgery for ovarian cancer? You could use donor eggs, but these are in short supply, given how gruelling the procedure is for collecting them. And in any case, you might want a child that is genetically related to you.

There may still be hope of that – not yet, but with luck in a

---

* The 14-day rule is adopted in law by several other countries, including Canada, Australia and Sweden. Some others, including the United States and China, make it a guideline, and it is advised in that sense also by the International Society for Stem Cell Research.

decade or two. It might become possible to make eggs "artificially" from induced pluripotent stem cells generated, as mine were, from a scrap of skin. In the embryo, some pluripotent stem cells will become the germ cells that give rise to gametes: eggs and sperm. We saw earlier that this is a specialized process, different from the formation of somatic cells in that the gametes have only one set of chromosomes and are generated in a bespoke mode of cell division called meiosis. All the same, there is no obvious reason why that should be beyond the capacity of induced pluripotent stem cells.

In fact, both "artificial" eggs and sperm might be created this way in the petri dish. And there's an urgent need for both. At least half of infertility problems arise because of poor sperm quality in the man, and it's getting worse: sperm counts in men fell by a staggering 50 to 60 per cent between 1973 and 2011. That doesn't imply a comparable drop in birth rates – couples in which the man has poor-quality sperm may simply have to try for longer before conceiving, or may elect to use donor sperm. But the issue with poor sperm isn't just about conception. Low sperm counts are often an indication of other health problems, actual or incipient, including testicular cancer, heart disease and obesity. Alarmingly, no one really knows what is causing the decline, although it seems likely that a mixture of environmental factors is to blame: poor diet, smoking, pollutants and chemicals that disrupt the development of the male reproductive system. At any rate, the decline in sperm count may be signalling a wider malaise in men's health.

Artificial sperm would not solve that broader problem, but it could alleviate infertility arising from low sperm numbers or quality. Already, IVF can work for men with low sperm count, because the sperm can be collected, concentrated and brought into contact with the egg. It can even work for sperm that can't "swim" – a common problem in low-quality sperm – using the ICSI technique to inject a sperm directly into an egg. But some men produce no sperm at all. In such cases, making sperm artificially by reprogramming stem cells could one day allow couples to have a biologically related child.

"When I think about it, it is quite astonishing," says Azim Surani, a developmental biologist at Cambridge who is one of the leading researchers in the production of artificial eggs and sperm. "Each cell in your body is a potential gamete. This is a profound change in the way we think about cells."

The notion of children "grown" from a piece of arm might sound bizarre, repugnant and even impious to some. There is of course a biblical resonance to this too, giving it the frisson of a god-like power:

> And the Lord God caused a deep sleep to fall upon Adam, and he slept: and he took one of his ribs, and closed up the flesh instead thereof;
>
> And the rib, which the Lord God had taken from man, made he a woman, and brought her unto the man.

But making gametes from stem cells is not as easy as making neurons. It requires a kind of recapitulation of the way natural gametes develop from embryonic stem cells. Some of those cells become selected for a germ-cell fate a few weeks after fertilization. First they form so-called primordial germ cells (PGCs), which move through the embryo until they reach the region developing into the gonads: the testes or ovaries.* It's only then that the gonads start to take on the features of one or other sex. If the embryo is male, some cells in the nascent sex organs produce a transcription factor from a gene on the Y chromosome denoted SRY (see page

---

* I'm being rather vague about exactly when and where the PGCs first form because, in humans, we simply don't know. It might be before or after gastrulation. And whereas the PGCs appear in the primitive streak in pigs, they are located elsewhere in monkeys.

71), which directs gonadal development towards testes. Otherwise they become ovaries by default.

Once in the gonads, the germ cells receive signals from the surrounding tissues that prompt them to mature into gametes. Meiotic cell division to form sperm happens constantly in males after puberty, but in females the diploid cells (with doubled chromosomes) called oocytes, from which haploid eggs are produced, begin meiosis in the fetus but then stay arrested halfway through that cycle for years, until the girl reaches puberty. They complete the cycle while lodged on follicles in the ovaries, before detaching to enter the fallopian tubes during ovulation. During meiosis, the epigenetic modifications that the chromosomes of PGCs have acquired are stripped away, resetting the genes to their pristine, pluripotent state.

So there's a lot to recapitulate *in vitro*: to turn stem cells into PGCs, then let them mature and undergo meiosis to form gametes. But it can be done – at least in mice. In 2011, biologists Mitinori Saitou of Kyoto University, Katsuhiko Hayashi of Kyushu University, and their colleagues created "artificial" sperm from the skin cells of adult mice after reprogramming them to iPSCs. They transformed the iPSCs into PGCs by injecting a single transcription factor, called BMP4.* They then transplanted these artificially induced PGCs into the testes of live mice, where the cells received the signals needed to complete their growth into sperm. The researchers used some of this sperm to fertilize mouse eggs, which developed into embryos

---

* The abbreviation stands for "bone morphogenetic protein 4", because this protein, and others in the BMP family, were first found to play a role in the formation of bone. But BMPs are now known to have a host of roles in development, some – as here – having nothing to do with skeletal growth. This is yet another indication that many genes don't have well-defined and easily summarized jobs, but induce different effects depending on when and where they are expressed as the body grows.

and then into apparently healthy mice pups. In 2016, a Chinese team claimed to have made artificial mouse sperm wholly *in vitro* and to have used it to fertilize eggs, transferring them into female mice for gestation. But some other scientists working in the field remain sceptical of those claims, which have not been repeated.

The analogous procedure can be used *mutatis mutandis* to make female gametes. Saitou's team has transplanted primordial germ cells, made in the same way from iPSCs or embryonic stem cells, into mouse ovaries, where the cells completed their development into eggs. The researchers have also developed a method for conducting the process entirely *in vitro*, using a kind of "artificial ovary" to provide the signals needed for complete maturation of the eggs, made from cultured mouse ovarian cells.

In one provocative experiment, Saitou's group took mouse eggs through an entire generational cycle without the intervention of adult mice at all. They made mouse eggs from pluripotent stem cells *in vitro*, then fertilized them by IVF using sperm from adult mice. They grew the resulting embryos to the blastocyst stage in a dish, at which point they could harvest new embryonic stem cells for the next round of gamete formation.

This experiment was a combination of procedures developed previously, yet in a sense it changes the whole nature of sex and reproduction. It means that one can grow a succession of mouse generations in a few days (the usual gestation period of a mouse is about 20 days) without ever making a single adult organism – indeed, without really making a "mouse" at all. Given a supply of sperm (which too can be made artificially if you wish), one can progress indefinitely from stem cell to egg to embryo to stem cell. The cells are, you might say, having sex and generating a genealogy without any need of the "organism" as such.

We don't really have words for what this procedure consists of. Biology, of course, doesn't care a whit about that.

This would all be dandy if there were any need to treat murine infertility (it's quite the opposite in my domestic experience). But what about making human gametes? We're some distance away from that. So far, researchers have coaxed human pluripotent stem cells to the stage of primordial germ cells, but getting them to mature further is tricky because again it requires the right signals from the gonadal cells. It turns out, though, that these signals aren't terribly species-specific. Mouse gonads can do the job – or at least part of it – for human germ cells too. In 2018, Saitou and colleagues reported that they had cajoled human PGCs onto the next stage of development towards eggs, called oogonia cells, *in vitro* by culturing them alongside mouse ovarian cells. Saitou admits that he thought it would be too much to expect the mouse tissue to have this effect on human PGCs, but he decided to try the experiment anyway. No, mice are not (wo)men – but as far as their cells are concerned, in this case they are close enough. It now remains to advance the oogonia on to become oocytes, and then to negotiate the challenge of meiosis that would make them into viable egg cells.

It seems likely that the analogous approach would work for advancing human PGCs towards sperm *in vitro*, by culturing them among mouse testicular cells. Again, whether fully mature sperm could be produced this way isn't clear, but it might not be necessary anyway: even somewhat immature sperm, lacking the swimming "tail", might be capable of fertilizing an egg if injected directly into it.

* * *

These are early days. We don't know for sure how closely the primordial germ cells or oogonia made artificially resemble those in human embryos or full-grown humans. We don't yet know how thorough their epigenetic resetting is: if the cells "remember" even a trace of their original lineage (as skin cells, say), they may fail to develop properly when fertilized. (The human oogonia-like cells made by Saitou's group do, however, seem to have been largely stripped of their epigenetic marks.) Some of the mouse eggs made from artificially induced PGCs look a little odd and misshaped, and they don't fertilize with the same success rate as "natural" eggs. In any event, no responsible researcher would countenance trying to use such cells for human reproduction without checking out safety questions like this in great detail.

All the same, artificial mouse gametes have yielded apparently healthy mice. And bioethicist Hank Greely of Stanford University says that he sees no obvious "show stopper" that might prevent what is feasible in mice ultimately from working in humans. Stem-cell biologist Werner Neuhausser of Harvard University says that "regeneration of human gametes from somatic cells in the lab is probably just a question of time and effort." The clinical need for such a technology, he adds, is tremendous.

Don't hold your breath, though. "I get lots of emails from people saying, 'my husband is fertile, he's desperate to have kids,'" says Surani. "Well, nothing is impossible, but this is very complex if you're going to think about clinical applications." To establish the feasibility and safety for human reproduction, he says there would first need to be work on non-human primates, which is slow, expensive and regulated almost beyond reach in some countries.

"Even with safeguards in place we would have to accept some residual risk," Neuhausser warns. "Ultimately, some patients would have to take a leap of faith if this technology enters clinical trials." Surani doubts that will happen within ten years, while Saitou's collaborator Katsuhiko Hayashi tends to warn people who contact him, eager to volunteer for IVF with artificial gametes, that using these methods for human conception could be as much as five decades off.

* * *

Reports of falling sperm counts fuel fears explored in dystopian fictions such as P. D. James's *The Children of Men* and Margaret Atwood's *The Handmaid's Tale*, in which human reproduction becomes almost vanishingly rare. Right now, there's no reason to imagine that such a scenario could come to pass. But it does seem that some aspects of modernity – whether it's changing diets and lifestyles or environmental pollution – could make fertility problems more widespread. And Richard Sharpe of Edinburgh University in Scotland, an expert on reproductive health, says that if the decline continues and the current lack of knowledge and research about causes and cures doesn't improve, it may one day become easier to use *in vitro*-generated male germ cells than naturally produced sperm to achieve conception. We might hope that we'll never need that option, but it's good to have an emergency plan. As bioethicist Ronald Green has put it, "If the human race as a whole were seriously endangered, and if our reproductive abilities were seriously compromised, we might have to manufacture human beings."

Yet beyond the rhetoric that artificial gametes will "end infertility" and "democratize reproduction", there are questions about how to think about fertility in the first place. Philosopher Anna Smajdor provocatively suggests that advances like these "could blow away the biological barriers of reproduction" – for example, making it possible for post-menopausal (even positively elderly) women and prepubescent children – indeed, as we saw above, even for embryos. Greely confesses that, despite having studied this field for many years, he has more than once been brought up short by suggestions of how these technologies might be used. One possibility is the "uniparent", whereby a person (either male or female) has both eggs and sperm made from their somatic cells and used to create a child (a "unibaby") – who would, because of the recombination of chromosomes during conception, not then be a clone in the strict sense.* Another is "multiplex parenting", where three or more people mix their genes to have a baby. In effect, says Greely, this could mean "two people want their child to mate with someone else, without the wait and bother of actually having a child and raising him or her to puberty."

The ability to make gametes from any bodily residue we leave lying around, like cells you leave on beer bottles and wine glasses, opens up other alarming scenarios. You can imagine the celebrity paternity suits already.

Such ideas, Greely concludes, are "evidence of just how wide-ranging and non-intuitive the implications of new biological technologies may be for human reproduction." Even the imagination of experts is boggled by the possibilities.

I imagine most people will find some of these scenarios grotesque. But, safety issues aside, the ethical questions aren't as straightforward as you might think, not least because no philosopher has yet resolved the rights and wrongs of bringing someone into existence versus their not existing at all. (How are a person's rights respected by denying them existence?) One thing seems clear: attempts to reach absolute judgements about rights and wrongs are likely just to hamper serious debate, while being outstripped anyway by the latest scientific advances. The guiding principle should surely not be "what?" but "why?" – with the "why" being directed towards the welfare of a child born this way. Smajdor's point is not that we should rush to embrace all options that arise or are imaginable. Rather, we are going to have to grapple with some hard questions about what we truly mean by fertility, reproduction and sex, and how we feel these things are – and should be – related.

---

* You could call it the ultimate form of incest – and it would incur all the same hazards of inbreeding that have made that a near-universal taboo.

CHAPTER 7

HIDEOUS PROGENY?

THE FUTURES OF GROWING HUMANS

Conversations about how to make a human often begin at the Villa Diodati on the shore of Lake Geneva in the "wet, ungenial" summer of 1816, where the teenaged Mary Godwin was trying to get to sleep. Some days earlier Lord Byron, the glamorous sovereign of that select little coterie of Romantics, had proposed that they each write a ghost story. And on every day that passed, Mary's friends had been entreating her: "Have you thought of a story?"

On that fateful evening, Byron and Mary's lover Percy Bysshe Shelley had discussed "the nature of the principle of life, and whether there was any probability of its ever being discovered and communicated." Perhaps, they had mused, "a corpse would be re-animated."

Night fell; Mary retired to bed but could not sleep. And then it came to her:

I saw – with shut eyes, but acute mental vision, – I saw the pale student of unhallowed arts kneeling beside the thing he had put together. I saw the hideous phantasm of a man stretched out, and then, on the workings of some powerful engine, show signs of life, and stir with an uneasy, half vital motion.

This, from the introduction that Mary – now Shelley, and widowed for almost a decade – wrote for the revised 1831 edition of *Frankenstein*, is as good as the account she gave in the book itself of how Frankenstein made his monster. "He sleeps," she went on, describing her nocturnal vision, "but he is awakened; he opens his eyes; behold the horrid thing stands at his bedside, opening his curtains, and looking at him with yellow, watery, but speculative eyes."

The moral of this tale, surely, was clear: "supremely frightful would be the effect of any human endeavour to mock the stupendous mechanism of the Creator of the world." *Frankenstein* showed what would happen when man tried to play God.

That is still how Mary Shelley's extraordinary book is usually interpreted today, and while rather few people now understand the accusation of "playing God" to involve genuine impiety, it serves as a shorthand admonition against the scientific hubris that attaches to attempts to manipulate and perhaps even to create life.

But what is most telling about this narrative is that society, not Shelley, has imposed it on *Frankenstein*. You might wonder why I say that, if the author herself stated that this is the book's message. But there's good reason to think that Shelley's 1831 introduction was shaped by the way society wanted to read her story, first published anonymously in 1818. In the earlier version, there were fewer references to Victor Frankenstein having tempted or challenged God through his grim work, and none of the original reviews raised that notion. Indeed, Percy Shelley seems much closer to the mark in the account of the book (which he helped to shape and edit*) that he wrote in 1817, published posthumously in *The*

---

* Percy was widely assumed to be the author when the book was first published.

Athenaeum in 1832: "In this the direct moral of the book consists: . . . Treat a person ill and he will become wicked." *Frankenstein* does not, in truth, have a moral that can be put into a sentence – but if we were forced to choose one theme, it ought to be this: we must take responsibility for what we create. For the life we create.

Yet by 1831, the Faustian interpretation of *Frankenstein* had become firmly established within the public consciousness, not least because many people had come across the tale via the bowdlerized, simplistic and hugely popular stage adaptations that began to appear in the mid-1820s. Mary Shelley, now with a public image to curate (her livelihood depended on it), and perhaps also touched by the conservatism that, alas, often arrives with age, responded by bringing her text into alignment with general opinion. It's possible too that she wanted to distance herself from the views of her husband's former physician William Lawrence, who had suffered condemnation from the Royal College of Surgeons for his materialist views of life: that it is "mere matter", requiring no mysterious animating soul. The influence of Lawrence's ideas can be discerned in the 1818 text; but his book *Lectures on Physiology, Zoology and the Natural History of Man*, published a year later, was fiercely denounced as irreligious and immoral. The changes that Mary Shelley wrought in the 1831 edition, making Victor Frankenstein more religious and pruning his scientific background, were, in the view of literary critic Marilyn Butler, "acts of damage-limitation".

The idea that *Frankenstein* is the foundational text warning against scientific hubris is, then, largely a twentieth-century view. This doesn't mean that we have got *Frankenstein* "wrong" – or at least, not simply that. It means rather that we needed a cautionary tale

---

Some others have suggested that he had as much to do with it as Mary, but that's not true. His edits are ornamental: little modifications of style, no more. And in any case, the roughness of Mary's prose and plotting is ultimately *Frankenstein's* gain, giving it the space in which myth can flourish.

to deal with our confusions and anxieties about life and how to make and change it – and *Frankenstein* can be interpreted to fit that bill.

The Faustian moral about over-reaching science might be, and indeed has been, applied to any technology, from nuclear energy to the internet. But *Frankenstein* is gloriously fleshy; that's what made it so repugnant to some of its first readers. The creature is created from flesh *as it should not be*: flesh made monstrous, flesh out of place. The review in the *Edinburgh Magazine* in 1818 entreated the author and "his" ilk to "rather study the established order of nature as it appears, both in the world of matter and of mind, than continue to revolt our feelings by hazardous innovations in either of these departments."

Shelley's novel showed human tissue given new forms, assembled into a frame of "gigantic stature". It is precisely this disturbing vision that we see revitalized, as it were, by the advent of tissue culture, with its tales of uncontrolled growth of human matter, of brains in jars and organs kept alive by blood controlled by pumps and valves, of chemical babies bred in vats.

These are the outer reaches of our new-found ability to grow and transform cells and tissues. At times, they do sound like what now is commonly invoked by the phrase "Frankenstein science". But they testify in favour of William Lawrence's materialist view of living matter, suggesting that the organism is indeed a collaboration of physiological structures, a "great system of organization" with no need of an intangible soul or spirit. We don't yet know how far the limits of that "organization" can be tested and extended, and neither are we quite sure where in that space of possibilities any concepts of humanness and self are located and corralled. But we are finding out. In doing so, we are not playing God – the phrase dissolves on inspection (theologically as well as logically) into a vague expression of unease and distaste. But we must never lose sight of the duty of care – to ourselves, our creations, and society – that Victor Frankenstein so woefully neglected.

* * *

Stitching or plumbing together tissues and organs seems an odd, clumsy and improbable way of making a human. For William Lawrence, life was intimately bound up with the animal body intact and complete. But by the early twentieth century, it no longer seemed absurd to consider the parts to be imbued with their own vitality. We've seen how cell theory was central to that reappraisal: many physiologists and anatomists were content by the turn of the century to regard the cell as an autonomous entity, independently alive. Tissue culturing vindicated that idea, as well as suggesting that flesh as an abstract entity could endure without the body's organizing influence.

Karel Čapek's *R.U.R.* shows how far-reaching these scientific ideas had become. As well as being a parable on the dehumanizing effects of the modern industrial economy, it synthesizes a great deal of the evolving thoughts about the nature of living matter. Yet since God was now officially pronounced dead, Čapek could risk invoking his name in satire. I do not know quite when the phrase "playing God" is first recorded, but it may very well be in *R.U.R.* Here is the company's director general Harry Domin discussing the founder's early work with the idealistic heroine who is visiting the factory to argue that the robots should be granted moral rights:*

Helena: They do say that man was created by God.

Domin: So much the worse for them. God had no idea about modern technology. Would you believe that young Rossum, when he was alive, was playing at God.

---

* Yes, the play is prescient in many ways; robot ethics is now a very active field.

It's perhaps easier to imagine the piecemeal assembly of a person (which is how R.U.R.'s robots are made) happening by degrees – by default, as it were – as body parts are replaced or renewed. This was what Alexis Carrel seemed to envisage for his perfused organs. They would confer a kind of immortality through continual renewal: each time an organ gives out, it is replaced by a new one. Presumably all that needs to be retained for continuity of the individual is the original brain and nervous system. When Carrel and Lindbergh wrote of "the culture of whole organs" in 1935, a newspaper headline insisted that they had taken the human race "one step nearer to immortality".

Through such constant renewal of parts, a person might become like Theseus's legendary ship, replaced little by little until no component of the original remains. Is it then still the same person? Edgar Allan Poe's story "The Man Who Was Used Up" depicts a process of that nature, albeit with replacement body parts that are mechanical rather than fleshy. They are used to patch up an old general injured in combat over his career – but as he becomes ever more mechanical, so too do his thought and speech. The scenario is played largely for comic effect, although it betrays a long-standing suspicion that turns out to have some element of scientific truth: the body does indeed affect the operation of the brain, and not just vice versa.

Of course, the brain too suffers wear and tear. Indeed, it may be the life-limiting organ: as average lifespans have increased over the course of the last century, we have become more prone to the debilitating and ultimately fatal neurodegenerative diseases that threaten the ageing brain, such as Alzheimer's and Parkinson's. Can the brain itself be replaced?

This was rumoured to be on Carrel and Lindbergh's agenda. As the two men met on Carrel's private island of Saint-Gildas, off the coast of Brittany, in 1937, reporters frustrated by lack of access came up with the most fanciful ideas. The *Sunday Express* claimed that the duo were engaged in "Lonely island experiments with machines that keep a brain alive".

Perhaps it's not surprising that such fantasies were common currency just six years after James Whale had shown "Henry" Frankenstein insert a bottled brain into his creature, as portrayed by Boris Karloff in the Hollywood movie. But the media are no less credulous today. Notwithstanding attention-grabbing headlines in 2017 of a successful human "head transplant" by controversial Italian neurosurgeon Sergio Canavero, no such thing has ever been achieved in any meaningful sense of the term, and it is far from clear if it ever could be. Canavero has previously claimed to have transplanted the head of a monkey, but the "patient" never recovered consciousness. Moreover, the surgeon did not reconnect the spinal cord (no one knows how to do that), so the animal would have been paralysed anyway. Canavero's claims for humans were based on experiments on dead bodies, and were a mere surgical exercise in connecting blood vessels and nerves, with no indication (how could there be?) that this restored function.

Besides, it's not really correct to call the procedure a "head transplant" even in principle. Imagine (even if it is highly unlikely) that a head could retain not just vitality but memories if separated from the body and kept "alive" by perfusion. And suppose that these memories could indeed be reawakened when the head was attached to a new body. Then it seems reasonable to assert that the brain, not the body, is what defines the patchwork individual so created – in which case this would more properly be called a body transplant.

But even if the thoughts and memories of a human brain can somehow be preserved in the face of the mortality of the flesh (I'll return to that shortly), we can safely say that, unlike say the heart, kidneys or pancreas, there is no current evidence that the whole human brain is in any sense a replaceable organ.

* * *

Visions of a full-grown human body assembled from component parts, from *Frankenstein* to Čapek and Carrel, remain fixated on the old Cartesian picture of body as mechanism. Sure, the parts are soft and fleshy, but they are otherwise viewed here rather like the cogs and levers of the ingenious automata of the Enlightenment, which were clothed in a mere patina of skin much like the intelligent robots of the *Terminator* films and Eva from Alex Garland's 2014 movie *Ex Machina*. In Carrel's time there was still an immense scientific and conceptual gap between studies of cells and tissues, which had begun to reveal their mysterious self-assembling autonomy, and the gross anatomy of the human body. We had little idea how one becomes the other.

It's different now; research on organoids brings that home very plainly. As these methods of cellular self-organization become better understood, it has become feasible to imagine and indeed to conduct experiments that shrink a piecemeal assembly process down to the microscopic scale: to build a human from an artificial aggregate of cells that can be considered a sort of synthetic, "hand-crafted" embryo.

Do we, in fact, need anything more than a single cell? That possibility was recognized when the advent of IVF began to reconfigure the way people thought about conception and embryogenesis. In 1969, Albert Rosenfeld wrote in *Life* that "there is the distinct possibility of raising people without using sperm or egg at all":

> Could people be grown, for example, in tissue culture? In a full-grown, mature organism, every normal cell has within itself all the genetic data transmitted by the original fertilized egg cell. There appears, therefore, to be no theoretical reason why a means might not be devised to make all of the cell's genetic data accessible. And when that happens, should it not eventually become possible to grow the individual all over again from any cell taken from anywhere in the body?

What Rosenfeld describes here is more or less identical to the process of turning a somatic cell into an induced pluripotent stem cell, making its full genetic potential "accessible" again.

This is, indeed, what iPSCs now make possible. In 2009, Kristin Baldwin of the Scripps Research Institute in La Jolla, California and her co-workers made full-grown mice from the skin cells (fibroblasts) of other mice. They reprogrammed the cells using the standard mixture of four transcription factors devised by Shinya Yamanaka, and then injected these iPSCs into a mouse blastocyst embryo. The blastocyst was prepared in such a way that its own cells each had four rather than the normal two chromosomes; that's done by making the cells of a normal two-cell embryo fuse together. Because these cells contain too many chromosomes, such embryos generally can't develop much beyond the blastocyst stage. But the added iPSCs didn't have that defect: they were normal two-chromosome cells. So the mouse embryo and fetus that developed from this blastocyst was derived just from the iPSCs.* Those fetuses grew to full-term pups which were delivered by caesarean section, and about half of them survived and grew into adult mice with no apparent abnormalities.

Now, not all iPSCs are capable of this: not all appear to be truly pluripotent, able to grow an entire organism. Some are and some aren't, and it's not entirely clear what makes the difference or how it can be spotted. But the experiment shows that at least some iPSCs are capable of becoming whole new organisms. Although this work was done with mice, there's no obvious reason why the conclusions would not apply equally to human cells.

Do you see what this means? It may be that pretty much every cell in your body can be grown into another human being. Had

---

* The Scripps researchers in fact used several different iPSC lines to make each embryo, each with distinct genomes – so the resulting mice were chimeras.

the iPSCs from which my mini-brain was grown been placed instead inside a human blastocyst, they could probably have been incorporated into, and perhaps have been entirely responsible for, a human fetus.

At this stage an experiment like that,* with all the attendant unknown and health risks, would be deeply unethical, and in some countries illegal. But I'm not recommending it; I'm simply saying that it can be imagined.

You might say that this thought experiment is not, however, quite a matter of growing iPSCs from scratch into a person. And you'd be right. You need the vehicle of a blastocyst to supply the crucial non-embryonic tissues needed for proper *in utero* development: the placenta and the yolk sac. Can we, though, add those too? Could we assemble clumps of cells and proto-tissues into what we might figuratively (some will say luridly) call "Frankenstein's embryo"?

No researchers currently feel it is either possible or desirable to actually make a human this way. The purpose of building an "artificial embryo" like this – or at least, a structure roughly resembling an embryo – is for fundamental research. It can supply a means of studying early human development that is not possible, or anyhow not permitted, with real embryos. In this way, we might learn more about what goes wrong – why so many embryos spontaneously abort – as well as what goes right in the *in utero* growth of humans.

Now that methods exist for culturing embryos at least right up to the 14-day limit legally imposed in several countries (see page 236), our relative ignorance of what directs embryo growth in the second and third weeks of development has started to look more

---

* I'm talking here about an experiment with reproductive goals. Creating a blastocyst that contains human iPSCs (which would be a chimeric embryo) for research within the 14-day limit is quite another matter.

glaring. "Synthetic embryos" or embryoids offer an alternative way of gathering knowledge about this critical stage of growth that doesn't conflict with ethical and legal restrictions. An added attraction of embryoids as a tool for basic research is that they might be tailored to suit the problem under investigation: refined, say, to supply a realistic model of how germ cells form or how the gut takes shape. You can keep parts of the picture quite vague and sketchy and just put in the detail you're interested in.

Embryoids have so far generally been made from embryonic stem (ES) cells rather than iPSCs. It has been known for over a decade that these cells alone can become somewhat embryo-like: in the right kind of culture medium, small clusters of them will spontaneously differentiate to form the three-layer structure that precedes gastrulation: the ectoderm (the progenitor of skin, brain and nerves), mesoderm (blood, heart, kidneys, muscle and other tissues) and endoderm (gut). There, however, the process typically stops: with a simple ball of concentrically layered cell types, endoderm outermost. In a normally developing human embryo, this triple layer goes on to develop the primitive streak and make the gastrula – the first appearance of a genuine body plan. For that to happen, the embryo needs to be implanted in the uterus wall.

In 2012, a team of researchers in Austria showed that implantation can be mimicked in a crude way by letting embryoid bodies made from mouse ES cells settle onto a surface coated with collagen, which can stand proxy for the uterus wall. Once the embryoid bodies are attached, their concentric, layered structure changes to a shape with bilaterial symmetry: with a kind of front and back. What's more, some of the cells start to develop into heart muscle. The embryoid begins to look a little more like a true *in utero* embryo.

Still, the cells can't be fooled for long. They won't continue organizing themselves unless they get all the right signals from their surroundings. Pretty soon they discern that there's no real uterus wall after all, and no placenta either. So they end up looking like a very crude gastrulated embryo – an infant's abortive attempt to make a figure out of clay, the features barely recognizable.

Here is where it pays to do a bit of assembly by hand: to add the tissues that the embryoids require if they are going to develop further.

The simplest recipe involves just two types of cell: pluripotent ES cells, and the cells that give rise to placenta, called trophoblast cells. It's these pre-placental cells that deliver key signals to ES cells *in utero* inducing them to take on the shape of a pre-gastrulated embryo. In 2017, Magdalena Zernicka-Goetz and her colleagues in Cambridge used this two-component recipe to create a mouse embryoid. Normal *in utero* embryos have a third type of cell too: the primitive endoderm cells, which form the yolk sac and supply signalling molecules needed to trigger the formation of the central nervous system. But as a first step, Zernicka-Goetz and her colleagues figured that the gel they used as the culture medium, infused with nutrients, might act as a crude substitute for the primitive endoderm: a kind of scaffold that would hold the embryoid in place as the trophoblasts did their job. They found that a blob of ES and trophoblast cells organize themselves into in a kind of peanut shape, one lobe composed of a placenta-like mass and the other being the embryo proper. After a few days, the embryonic cells developed into a hollow shape, the central void mimicking the amniotic cavity that forms in a normal embryo.

Zernicka-Goetz and her colleagues found that the patterning in these embryoids is dictated by the same genes that are responsible in actual embryos: for example, the morphogen called Nodal, which creates a kind of body axis around which the pseudo-amniotic cavity opens up, and the morphogen BMP, which triggers the emergence of the primordial-germ-cell-like cells. It's just like real embryogenesis: a progressive unfolding in which each step of the patterning process literally builds on the last. Get one stage of the shaping right, and you produce the conditions to turn on another set of morphogen genes, creating the chemical gradients for the next step of the process.

What's more, the researchers found that some of the ES cells in the embryoid start to differentiate towards particular tissues – not just the mesoderm that forms many of the internal organs but also cells that look rather like primordial germ cells, which go on to form sperm or eggs.* That's particularly exciting, because it raises the prospect of being able to make artificial gametes (see page 238) not by painstaking reprogramming of isolated stem cells but by housing them in a synthetic embryo-like body, which might be expected to make the process easier and more like that in normal embryos. Making fully functional germ cells this way, however, is still a speculative idea.

There's a limit to how far you can get this way. Sooner or later, the embryoid is going to figure out that it's not in a real womb but just a blob of gel. At that point, it will metaphorically stamp its foot and say, I'm stopping right here.

No one knows how long you can postpone this moment of protest. But it can be delayed by supplying the missing primitive endoderm cells themselves as part of the recipe, rather than just mimicking them crudely with the gel matrix. Zernicka-Goetz and her colleagues find that such three-component embryoids, which they make as

---

* I'm making the claim rather casually that this or that cell becomes a particular type: mesoderm, primordial germ cell and so on. How do we know, given that most cells look largely alike at this stage under the microscope? Researchers use the same methods that Chris Lovejoy and Selina Wray did to look at the cell types in my mini-brain. They deploy molecular tags that stick to particular proteins and emit fluorescent light when illuminated. Certain of these proteins serve as fingerprints of the cell type: they are known to be made only by particular cell types. In this way, these cell structures become gorgeous kaleidoscopic patterns of coloured blobs, showing exactly which types of cell are present, and where.

clumps of the respective cells floating freely in a nutrient solution, truly resemble embryos entering the process of gastrulation. One of the key signatures of that process is that the layers of cells around the "amniotic" cavity turn into so-called mesenchymal stem cells, which in a normal embryo will move about and form the mesoderm: the progenitor cells of the internal organs. The three-component embryoids show these characteristic steps of a transition to mesenchymal cells and formation of mesoderm.

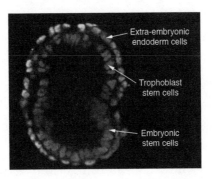

A mouse embryoid made by assembling embryonic stem cells
with two types of extra-embryonic cell: the trophoblasts that would
form a placenta, and extra-embryonic endoderm to make a yolk sac.
The cells organize themselves into an embryo-like structure
with an internal cavity.

There's no obvious reason why a patchwork embryoid of this sort, given all the right cells and signals, should not keep on growing and reshaping itself further. If we could prolong the deception, as it were, that an embryoid is *in utero*, could it get to something like a fetal stage? Might we, alternatively, assemble cells from scratch into an embryoid that looks enough like a pre-implantation blastocyst to be implanted and grown in a uterus, making a person not only without sexual intercourse but without even sperm or egg?

Well, bear in mind that most of the embryoid work I've described so far used mouse cells. Many of the basic processes in early mouse embryogenesis are the same as those in humans, but development becomes very different even at the stage of gastrulation: the mouse gastrula looks quite unlike the human one. Yet it's not obvious that there should be any fundamental obstacle to making human embryoids of at least this level of complexity. Human trophoblast cells – the vital ingredient for getting a placenta-like signal – aren't easy to make, but that was achieved in 2018. What's more, human trophoblasts have been grown into organoids that mimic a real placenta, raising the possibility of an *in vitro* tissue that can nurture embryoids as a stand-in for the maternal environment. And a team led by Jianping Fu at the University of Michigan has already grown human embryonic stem cells by themselves into embryoids that can recapitulate some of the stages involved in the formation of the amniotic sac.

We saw earlier that mere clumps of human ES cells can, in the right conditions and given the right biochemical signals, develop layered structures reminiscent of those in a pre-gastrulated embryo. With a bit more help, they can get even further. Ali Brivanlou of the Rockefeller University and his collaborators have shown that, if these rather crude human embryoids are exposed to a morphogen protein called Wnt, first identified in fruit-fly embryology (the W stands for "wingless", describing a mutation it can cause), they can grow into a shape resembling a primitive streak, marking the onset of gastrulation.

That's very striking. We saw that the development of a primitive streak in the *in utero* human embryo is taken as a rough-and-ready signature of "personhood" for the purposes of regulation: it's the basis of the 14-day rule. Does a feature like this in a "synthetic" embryoid qualify it for any comparable status? But there's more: Brivanlou and colleagues found that a combination of Wnt and another protein called Activin takes matters still further: it can induce in the embryoid the appearance of a group of "organizer" cells, of the kind that define the key axes for body formation. To show that this is genuinely an organizer, the researchers grafted the human embryoids onto chick embryos, and saw that the embryoid influences the growth of the chick embryo by introducing a new axis of development and triggering the appearance of progenitor cells to neurons – just as one would expect for an organizer that orchestrates the formation of the central nervous system.* It really does seem, then, as though these manipulations of human embryonic stem cells are reproducing some of the key early stages in the emergence of the body plan. Such results lend credence to the claim by stem-cell biologist Martin Pera of the Jackson Laboratory in Maine that "there is no reason to believe that there are any insurmountable barriers to the creation of cell culture entities that resemble the human post-implantation embryo in vitro."

Perhaps the barriers are not technical but conceptual. We need to ask: what sort of beings are these?

* * *

Embryoids are not exactly synthetic versions of the equivalent structures in normal embryogenesis. Their circumstances are different, and therefore so are some of their fine details. Like my mini-brain, they are both like and unlike the "real thing". None yet has the slightest potential to continue its growth *in vitro* towards a baby animal. They are a class of living things in their own right. Anticipating that the very rudimentary embryoid structures so far created with human cells will evolve towards the more advanced

---

* In other words, signals from the *human* cells are governing development in the *chick* embryo. Here's another reminder that, at the cell level and at this early stage of embryo growth, the apparently major differences between species don't much matter: the key genes and proteins involved are much the same in humans and chickens.

forms made from mouse cells, synthetic biologist George Church of Harvard University has proposed that we call this family of existing and prospective living objects "synthetic human entities with embryo-like features", or SHEEFs.

It may or may not catch on, but the rather cumbersome term summarizes what matters. The trajectory of this book has been precisely towards a conjunction of those first three words. These structures are *synthetic* because nature alone would not produce them. They are *human* to the extent that all the cells have human DNA. And they are *entities* because . . . well, because it is hard to find another word to lodge between the cell and the fully developed organism, granting individuality while remaining agnostic about selfhood.

It's true, albeit a rather banal truth, that SHEEFs sound like something out of *Amazing Stories*. Yet their possibility has been staring us in the face ever since we first granted cells the status of autonomous living things. We humans are a particular arrangement that those living things can adopt. A very special arrangement, to be sure, because it is the one evolution has conferred on the totipotent human cell in its normal environment of division and growth. But there is no reason to think it is unique. And embryoids don't have to fully recapitulate normal development, or adopt the same shapes as normal embryos, in order to be useful research tools.

Most researchers in the field recommend a ban on the use of embryo-like entities made from stem cells for reproductive purposes. But even if this were not to mean "growing a human", research on ordinary human embryos is much more tightly constrained than that – in particular by the 14-day rule observed in many countries. It's not clear, though, how or if that might apply to embryoids, given that they have no obligation to follow a natural developmental pathway. The 14-day limit is dictated by the appearance of the primitive streak, which will eventually become the central nervous system. But an embryoid might be tailored so that it reproduces some aspect of development normally evident only later in embryogenesis, without having to make a primitive streak first. It might *never* develop a primitive streak. But if so, how can we decide on a time limit for its growth?

These entities alter our view of the entire developmental process. Previously, it was seen as a single highway, with checkpoints all travellers must pass. But as Church and his colleagues have argued, SHEEFs change the highway to a network of many paths. We can take new routes, perhaps avoiding the conventional routemarkers altogether, and perhaps reaching new destinations. No one knows quite what is possible.

Partly for that reason, there is no consensus on how to legislate research on embryoids and SHEEFs. It's not just that their status is ambiguous; there is no standard form for an embryoid, any more than there is for an aeroplane or a television. They are put together however we want them, and the cells work with what they are given. "It is unclear at which point a partial model [of the embryo] contains enough material to ethically represent the whole," said a group of experts in this field at the end of 2018.

Throw into this pot the option of genome editing – which I will come to shortly – and the possibilities become dizzying, some might say terrifying. What if we found a way to make a human embryoid lacking the genes to make a proper brain, but which could nevertherless develop into a fetal body containing nascent organs that might be used for transplantation? Or, supposing some low-level brain functions to be needed for effective maintenance of the body, maybe we could engineer an embryoid with a "minimal" brain, lacking the regions and functions needed to feel pain or sentience? These are fanciful visions, but they are not obviously absurd. More to the point, they illustrate how hard it might become to develop clear moral and ethical reasoning about embryoids and SHEEFs. We can probably agree that SHEEFs able to register pain or sentience cross some moral boundary. But if they lack that capacity, are they really moral beings at all, even if made of human cells and vaguely humanoid in shape? I can't articulate a logical argument for why it would be morally wrong to create such entities, but the idea makes me uneasy to say the least. Would it make things any better if they were designed *not* to look troublingly humanoid? There's an urgent need to start the discussion now, because the science is likely to catch up sooner than we might imagine.

* * *

Even some of the more accessible and, from some angles, rather innocuous reproductive technologies that IVF has made possible are unsettling traditional notions of what is entailed in making humans. Take mitochondrial replacement. Mitochondria, we saw earlier, are organelles in our cells that generate the energy-rich molecules used in most enzymatic processes – they are, if you like, the furnaces of metabolism. Uniquely among the cell's compartments, they have their own dedicated genes, separate from those on the 23 chromosomes; that's one of the reasons to believe that the mitochondria are relict forms of what were once separate single-celled organisms (see page 97). Some mutant forms of the mitochondrial genes can give rise to so-called mitochondrial diseases, which can be highly debilitating and even life-threatening.* As mitochondria are inherited only from the mother, diseases arising from defective mitochondrial genes can be passed on down the maternal line.

---

* By no means all mitochondrial diseases with genetic roots are caused by the mitochondrial genes themselves, however; around 85 per cent of them are caused by mutations in the nuclear DNA.

Mitochondrial replacement therapies aim to avoid inheritance of these kinds of mitochondrial disease. There are various approaches, but all involve transplanting the chromosomes of an affected egg – perhaps after being fertilized by the biological father – into an egg donated by a woman with healthy mitochondria from which the chromosomes have been removed. If this zygote then develops into an embryo, its genes all come from the biological mother and father except for those in the mitochondria, which come from the egg donor. Insofar as it involves the transfer of chromosomes into a "de-chromosomed" egg, the technique uses the same kind of manip-ulation of cells as in cloning by nuclear transfer (see page 151).

Mitochondrial replacement has become possible thanks to IVF: the manipulations, including fertilization, all happen in a petri dish. Alison Murdoch at Newcastle University, whose group has pioneered the technique, says that "if we had not had the prior experience in [*in vitro*] human embryology . . . we would probably never have been able to achieve this."

The treatment has been highly controversial, partly because of questions about long-term safety that are almost impossible to answer with certainty for such a new medical technique. Replacing the mitochondria also amounts to a genetic alteration of the "germ line": if the embryo is female then this genetic change will be transmitted to any future female offspring. Even though it's hard to see what objection there could be to eliminating a thoroughly nasty disease throughout the subsequent generations, there are good reasons for scientists to proceed very cautiously with any genetic alteration that affects the germ line. All the same, the method was approved for use in the UK by the Human Fertilisation and Embryology Authority (HFEA) at the end of 2016, and in early 2018 the authority granted a licence to a Newcastle hospital for use of the method in IVF for two women with mitochondrial disease mutations. An apparently healthy child has already been born from the technique in a clinic in Mexico, supervised by a doctor from a fertility clinic in New York.

Some of the opposition to mitochondrial replacement has been driven not by concerns about safety or the wisdom of germ line alteration but by accusations of "unnaturalness". Because the genetic material in babies conceived this way would have a small component – the 37 genes in the mitochondria – that comes from the egg donor rather than the biological mother and father, such infants have been dubbed "three-parent babies". Everything that feels unsettling about new reproductive technologies is crystallized in that term: it suggests a runaway can-do science that has far outstripped our conventional moral boundaries and categories, permitting biological permutations that our instincts tell us should not be possible. By permitting such creations, critics say, we aren't just playing God but exceeding the bounds of nature, tinkering with the natural order "just because we can".

It's a perfect illustration of the use of narrative to frame – here in both senses – the science. "Parent" joins a list of words like "sex", "person" and "life" that have strong cultural resonance while being rather hazy in any scientific sense. All the emotive power of "parenthood" is here being channelled onto a handful of genes, in order to embed a biological procedure within a particular story, for a particular agenda.

For the fact is that when parenthood, with all the associations of kinship, responsibility and nurture, becomes reduced to a tiny strand of DNA that orchestrates some of the biochemistry within a specific cell organelle, anyone who has ever parented a child* has cause to feel affronted. So too do the families in which there are already effectively three or more parents, whether through step-parenting,

---

* Including, I'd hope it goes without saying, a child with whom they share no genetic relationship.

same-sex relationships, adoption or whatever. Like the "test-tube baby", the "three-parent baby" is not a neutral description but a label designed to convey a particular moral message and invite a particular response. It's up to us whether we accept it.

* * *

Which bring us to perhaps the most notorious bogey, the Huxlerian worry-doll, of the age of IVF: the designer baby. As the technology was taking off in the late 1960s, *Time*'s medical writer David Rorvik described the future according to reproductive biologist E. S. E. Hafez:

> [He] foresees the day, perhaps only ten or fifteen years hence, when a wife can stroll through a special kind of market and select her baby from a wide selection of one-day-old frozen embryos, guaranteed free of all genetic defects and described, as to sex, eye colour, probable IQ, and so on, in details on the label. A colour picture of what the grown-up product is likely to resemble, he says, could also be included on the outside of the packet.

As ever, a glance back puts our present-day fears in perspective. It's remarkable both how little the popular, commercially inflected image of designer babies has changed over the past four decades and how little real progress there has been towards it.

Here's the premise. Since the 1970s, it has become possible to edit genomes: to excise or insert genes at will, sometimes from different species entirely. What then is to stop us from creating an IVF embryo and tampering with its genes to alter and enhance the traits of the resulting child? Might we tailor her (for we can certainly select the sex) to have flame-red hair and green eyes, to be smart and athletic, full of grace and musical ability? (The stereotypical wishlist for discussing designer babies is generally along lines like these.)

Let's first have a look at what is possible.

Genome editing* has become rather routine in biotechnology. Organisms modified this way – especially bacteria – are used industrially for producing useful chemicals and drugs. In one especially striking (though not very commercially important) example, goats have been genetically modified with the genes responsible for making spider silk so that they secrete the silk proteins in their milk.

In bacteria it's not necessary to modify the cell's genome; one can simply inject additional DNA with the requisite gene, which the cell's transcription machinery will process in the usual manner to produce the desired protein product. But editing the genomes of organisms like us to change or restore the function of a gene is generally a cut-and-paste job: excising one gene, or part thereof, and inserting another. This demands tools for snipping the DNA in the right place, removing the gene fragment, and splicing in the new piece of DNA.

Molecular scissors and splicers that work on DNA have existed for decades – there are natural enzymes capable of these things that can be adapted for the job. But a technique called CRISPR, developed in 2012 largely by biochemists Emmanuelle Charpentier, Jennifer Doudna and Feng Zhang, has transformed the field because of the accuracy with which it can target and edit the genome. CRISPR† exploits a family of natural DNA-snipping enzymes in bacteria called Cas proteins – usually one denoted Cas9, but others are finding specialized uses too – to target and edit genes. Bacteria have evolved these enzymes as a defence against pathogenic viruses:

---

* One sees "gene editing" and "genome editing" used interchangeably, and there is no hard-and-fast definition. It is possible in principle both to edit an individual gene by switching a few base pairs, perhaps turning a disease-causing variant into a healthy one, and to replace an entire gene or set of genes to the same end. Both are done. Alterations to genes that are intended to change the phenotype of a cell or organism – in effect, to change a trait – are best described as genome editing, acknowledging that very few genes act in isolation.

† There is little to be gained by spelling out the acronym, which refers to the

the enzymes can recognize and remove foreign DNA inserted by the viruses into bacterial genomes. The targeted section of DNA is recognized by a "guide RNA" molecule carried alongside Cas9: the sequence of base pairs on the DNA complements that of the RNA. In other words, the RNA programmes the DNA-snipping Cas9 enzyme to cut out a particular sequence – *any* sequence you choose to write into the RNA.

CRISPR is far more accurate, as well as cheaper, than previous gene-editing techniques. If it proves to be safe for use in humans, then gene therapies – techniques for eliminating and replacing mutant genes that cause some severe, mostly very rare diseases – might finally become possible after decades of rather fruitless effort. The method could potentially supply a powerful way to cure diseases that are caused by mutations of one or a few specific genes, such as muscular dystrophy and thalassemia. Clinical trials are now underway.

Gene-modified embryos and babies are another matter. Gene therapies aim to alter genes in somatic cells, but any changes to genes made in an embryo at the outset will be incorporated into the germ line and passed down future generations. As we saw earlier, scientists are rightly hesitant about introducing changes to the germ line. What's more, if the editing process makes any other inadvertent alterations to the genome at this early stage in development, those changes will be spread throughout the body as the embryo grows.

CRISPR has been used to genetically modify human embryos, purely to see if it is possible in principle. The first results, obtained in 2015 by a team in China, were mixed: editing worked, but not with total reliability or accuracy. The following year, Kathy Niakan of the Francis Crick Institute in London became the first (and so

---

types of DNA sequences that are involved in the natural gene-editing process in bacteria.

far the only) person to be granted a licence by the Human Fertilisation and Embryology Authority in the UK to use CRISPR on human embryos. Her group had no intention to modify embryos for reproductive purposes, which remains illegal in the UK. Rather, they studied embryos just a few days old to find out more about genetic problems in these early stages of development that can lead to miscarriage and other reproductive problems.

Then in 2017, an international team led by reproductive biologist Shoukhrat Mitalipov at the Oregon Health and Science University reported the use of the CRISPR-Cas9 system to snip out disease-related mutant forms of a gene called MYBPC3, which causes a heart muscle disorder, in single-celled human zygotes, and replace it with the healthy gene. They fertilized eggs with sperm carrying the mutant gene, injected the CRISPR system into the zygotes to replace the genes with the properly functional form, and analysed the resulting embryos at the four or eight-cell stage to see how well the gene transplant had succeeded. In most cases, the embryos had only the "healthy" gene variant.* Unlike the Chinese experiments, these ones involved embryos that could in principle continue to grow if implanted in a uterus, although that was never part of the plan.

Indeed, the use of CRISPR for human reproduction is forbidden in all countries that legislate on it, and is almost wholly rejected in principle by the medical research community. Even research on genetic modification of human embryos that does not have repro-ductive goals cannot legally receive federal funding in the United States, although privately funded research (like Mitalipov's work) is not banned. And in 2015, a team of scientists warned in *Nature* that genetic manipulation of the germ line by methods like CRISPR,

---

* Curiously, the Cas9 enzymes here were guided to the faulty MYBPC3 gene in the paternal DNA by RNA made from the normal gene on the maternal DNA, rather

even if focused initially on improving health, "could start us down a path towards non-therapeutic genetic enhancement".

Until the end of 2018, most researchers in this field felt confident that, given the risks and uncertainties of CRISPR genome editing in human reproduction, and the almost unanimous view among researchers that it should not be tried until much more was understood, if at all, there was no prospect of genome-edited babies any time soon. They were shocked and dismayed, then, when a Chinese biologist named He Jiankui affiliated to the Southern University of Science and Technology in Shenzhen announced at a press conference in Hong Kong in November that he had used the method to modify IVF embryos and implanted them in women, one of whom had already given birth to twins.

The story was mysterious, bizarre and disturbing. He said that he had used CRISPR to alter a gene called CCR5, which is involved in infection of cells by the AIDS virus HIV. If the gene is inactivated, that could hinder the virus from entering cells. His aim, he said, was to produce babies resistant to HIV, so that couples who carry the virus might have children without fear that AIDS would be transmitted to them. (After a history of denial, China now recognizes that AIDS is a problem of epidemic proportions.)

He Jiankui offered few details, but it seemed he had added the molecular machinery of the CRISPR-Cas9 system to the embryos after fertilization, and had removed cells from them a few days later for genetic testing to see if the CCR5 gene had been successfully modified. If so, the couples were offered the choice of using either edited or unedited embryos for implantation. He said that a total of 11 embryos were used for six implantations before a (non-identical twin) pregnancy was achieved. In the press

---

than by the guide strands that the researchers provided for it. They figured that the CRISPR process thus happens rather differently for embryos than for somatic cells.

conference, he merely stated that "society will decide what to do next".

He Jiankui provided no documented evidence to back up his claims, nor did he say where the work was done or with what funding. But although stem-cell and embryological research has been plagued with fraudulent claims, this one seemed genuine. He was an academic with a respectable pedigree, having studied in the United States at top universities. His adviser at Rice University in Houston, American bioengineer Michael Deem, admitted to collaborating on the work, though it was done in China.

The work was almost universally condemned as deeply unethical and unwise by specialists around the world. Over one hundred Chinese scientists put out a statement denouncing it and saying that it would damage China's reputation for responsible research in this field (a reputation that, contrary to what is sometimes asserted, is largely justified). "The experiment was heedless," wrote cardiologist Eric Topol of the Scripps Research Institute in California. "It had no scientific basis and must be considered unethical when balanced against the known and unknown risks." Jennifer Doudna, one of the inventors of the CRISPR technique, issued a statement saying, "It is imperative that the scientists responsible for this work fully explain their break from the global consensus that application of CRISPR-Cas9 for human germline editing should not proceed at the present time." He's university denied knowledge of the work (although it now seems possible that he used state funding to conduct it) and has now fired him, while authorities at Rice began an investigation into Deem's involvement. He's work is now under investigation by the Chinese authorities, and he could eventually face criminal charges if he is found to have subverted ethical rules or compromised the babies' health. Several leading scientists in the field have now called for "a global moratorium on all clinical uses of human germline editing".

In her statement, Doudna added that there was an "urgent need to confine the use of gene editing in human embryos to cases where a clear unmet medical need exists, and where no other medical approach is a viable option." That was not the case here: the embryos that He Jiankui treated had no known intrinsic genetic defect, but were modified *in anticipation* of HIV infection. It wasn't clear that the strategy would be effective anyway: one of the twins in the pregnancy apparently acquired only one copy of the altered CCR5 gene, and so would not get complete resistance to HIV. Besides, there are other ways of treating HIV infection. What's more, knocking out the CCR5 gene brings dangers in itself: a lack of the functioning gene increases susceptibility to some other viral infections. All in all, it was a barely explicable choice for a first use of CRISPR in human reproduction, quite aside from the unknown risks. To make matters worse, the work seemed to have been done shoddily: as more details emerged, it became clear that there had been some off-target genome modification, with unknown consequences.

The eyes of the world will be on these two children (both girls). They are the proverbial guinea pigs, facing an uncertain future due to the recklessness of maverick researchers.

As Doudna and others realized, this headline-grabbing and irresponsible act would only damage the prospects for potentially valuable uses of the technology. For there is no obvious reason why genome editing for human reproduction should be forever ruled out of court. In those relatively rare cases where a debilitating disease is caused by a single gene, and the consequences of replacing a faulty with a healthy version can be reliably predicted, there may be a place for it eventually in reproductive medicine – although it is far from clear that it would be a better option than alternative strategies for tackling the medical problems it would seek to address.

After all, as with mitochondrial diseases, eliminating such

diseases from the germ line seems an unqualified good. Why would you not want to ensure that not only a person grown from the modified embryo but also their offspring too are free of the disease? Some of the qualms arise from the irreversible nature of the change – what if it turned out that the excised gene mutant, as well as creating a disease risk, had some beneficial value, in the way that the gene variant that causes sickle-cell disease when on both chromosomes can confer some resistance to malaria when on just one? But in the unlikely event of a gene that causes some serious disease turning out to have unforeseen benefits too, genome editing is *not* irreversible: a mutant gene can be restored as well as removed. In February 2019 the World Health Organization established a committee to draw up guidelines for human genome editing, which will doubtless wrestle with questions like these.

The new techniques of cell reprogramming and repurposing expand the possibilities for genome editing. "Pretty much any genetic modification could be introduced in iPSC lines derived from somatic cells using CRISPR," says stem-cell scientist Werner Neuhausser. The generation of "artificial gametes" or even embryos from somatic cells would open doors in reproductive technologies because they could make the safety margins wider. If those methods created a ready supply of human eggs for IVF, one could carry out the editing procedure on many eggs or embryos at once and pick out instances where it worked well – there would be slack in the system to accommodate a bit of inefficiency and mistargeting. At any rate, bioethicist Ronald Green believes that human genome editing "will become one of the central foci of our social debates later in this century and in the century beyond." For better or worse, the headline-grabbing antics of He Jiankui may well have launched us on that trajectory sooner than anyone expected.

* * *

Would embryos genome-edited to prevent disease be "designer babies"? That would seem a perverse use of the pejorative label. It is supposed to convey a sense that the babies are luxury items – accessories that feed our vanity. But disease prevention is no luxury.

More often, the term is invoked in discussions about what we might call the positive, non-medical selection of traits – not eliminating undesirable genes that create health risks, but selecting ones that will make a person smarter, better looking and better performing.

But predicting traits like these from genes is far more complicated than is often assumed. Talk of "IQ genes", "gay genes" and "musical genes" has led to a widespread perception that there is a straightforward one-to-one relationship between our genes and our traits. In general, it's anything but. As we saw earlier, most of our traits, including aspects of personality and health, emerge from complex and still largely inscrutable interactions between many genes, most of which have negligible impact on that trait individually. Intelligence, as measured by current metrics such as IQ (and we can argue elsewhere about their merits), is significantly heritable, meaning that it has genetic roots. Typically around 50 to 70 per cent of a person's "intelligence" is thought to be due to genetic factors. But there are no "intelligence" genes in any meaningful sense; the many genes that influence intelligence doubtless serve other roles (often connected to brain development).

Scientists have got much better in recent years at being able to make predictions about intelligence from a person's genome sequence, largely because of the availability of more data from many thousands of individuals. Given enough data, it becomes possible to spot even very small correlations between genes and intelligence. But these predictions are and always will be probabilistic. They are along the lines of "your genome suggests that you're likely to be in the top 10 per cent of the population for IQ/school achievement/ exam results". But environment has an influence too: if a person with a "good genetic profile" has suffered extreme neglect or abuse, or has had a serious head injury, their IQ might be far lower than their genes predict. And even without considering the influence of environment, a given genetic profile for intelligence will have a wide range of outcomes because it only biases, and does not totally prescribe, the way the brain wires up. There is some random "noise" in the developmental programme.

It's the same for other behavioural traits – creativity, say, or perseverance, or propensity for violence. This goes for many of the more common diseases or medical predispositions with a genetic component too. While there are thousands of mostly rare and nasty genetic diseases that can be pinpointed to a specific gene mutation, most of the common ones, like diabetes, heart disease or certain types of cancer, are linked to several or even many genes, can't be predicted with any certainty, and depend also on environmental factors such as diet.

So the genetic basis of attributes like intelligence and musicality is too thinly spread throughout the genome to make them amenable to design by editing. You would need to tinker with hundreds, perhaps thousands of genes. Quite apart from the cost and practicality, such extensive rewriting of a genome would be likely to introduce many errors. And because all those genes have other roles, you couldn't be sure what other traits your putative little genius would end up with: he might be insufferable, sociopathic, lazy. "The creation of designer babies is not limited by technology, but by biology," says epidemiologist Cecile Janssens at Emory University in Atlanta. "The origins of common traits and diseases are too complex and intertwined to modify the DNA without introducing unwanted effects." And it would all be for the sake of a probabilistic outcome that might still turn out to be disappointing: the tail end of the bell curve for those with "top 10 per cent IQ" genes reaches well into the realm of mediocrity.

Unfortunately, it is the same with those "many-gene" diseases. Genome editing might help to eliminate a nasty single-gene disease like cystic fibrosis, but probably won't help much with an inherited propensity to heart disease.

Is the designer baby off the menu then? Not quite.

*   *   *

Genome editing would be a difficult, expensive and uncertain way to achieve what can mostly be achieved already in other ways. For the fact is that we are already in effect genetically modifying the germ line of babies – not by changing their genomes, but by selecting embryos with particular genes in preference to others.

At present, this is mostly done for what most would agree are good reasons: to avoid serious disease. Couples in which both partners know they are carriers of a gene variant that causes disease – that's to say, they have just one copy of the gene on their pairs of chromosomes* – may elect to have a child using the procedure called pre-implantation genetic diagnosis (PGD), which can spot if an embryo inherits the "bad gene variant" from both parents and so will develop the disease. (There is a one in four chance of that: of the four possible combinations of gene variants from the two parents, only one corresponds to a double copy of the "disease" variant.) In PGD, embryos produced by IVF are genetically screened by removing a cell at around the four or eight-cell stage and sequencing its genome. Only recently has it become possible to do this cheaply and quickly enough to make the technique possible. The screening process shows which embryos can be safely implanted.

PGD is already used in around 5 per cent of IVF cycles in the USA, and in the UK it is performed under licence from the HFEA

---

* Gene variants that cause single-gene diseases are often "recessive", which means to screen for around 250 genetic diseases – those in which a single gene is known to cause or present a high risk of incurring the disease, which include thalassemia, early-onset Alzheimer's and cystic fibrosis.

PGD is not generally thought of as "germ-line genome editing". But that's really just because no one elects to call it that. If you want to have a randomly selected grab-bag of sweets with no green ones in, you could either take a single bag and replace the green ones by hand or take a hundred bags and look for ones with no green sweets in them. The end result is the same.

Putting it another way: we already seem to accept that there is nothing wrong in principle with specifying some genetic aspects of the germ line. The discussion is really about what those aspects may and may not be.

For one thing, what counts as a disease? What would selection against genetic disabilities do for those people who have them? "They have a lot to be worried about here," says Hank Greely, "in terms of 'Society thinks I should never have been born', but also in terms of how much medical research there is into their disease, how well understood it is for practitioners and how much social support there is."

In the UK and some other countries, PGD remains strictly forbidden for any purposes other than avoidance of specified diseases like those listed on the HFEA's register. It would not be

---

that you'll only get the disease if you have the variant on both of the respective chromosomes – inherited from both parents. If you have one "disease" variant and one copy of the normal gene, the normal form is "dominant" and you don't suffer from the disease. Indeed, you might not ever know you are a carrier: it's thought that everyone is a carrier of some recessive disease-related gene varieties. But carriers of such genes sometimes do know it, because a relatively high risk of incurring the disease runs in the family. Some disease-related genes, though, are dominant – you need only one copy of the wrong variant to get the disease. The MYBPC3 gene edited out of embryos by CRISPR as described earlier is like this.

permitted, say, for selecting the sex or hair colour of a child. But around 16 countries, including the United States, *do* permit sex selection. In these cases, there seems to be no preference for either a boy or a girl. Still the practice is controversial, not least because there are good reasons to think that some cultures *would* express a strong preference (for boys). Because of traditional attitudes towards gender, abandonment, mistreatment and even infanticide of baby girls have seriously skewed the male-to-female ratios in China and India, despite efforts of governments to promote the equal value of both sexes. The worst affected regions in these countries now face the prospect of social unrest due to large numbers of young men with no marriage opportunities.

Where prejudices about the gender of a child do not seem to exist, however, sex selection for "family balancing" doesn't seem obviously abhorrent. One might argue that gender should be irrelevant to how a child is valued; but is it so unreasonable for the parents of three boys to wish for a girl next time? Advocates point out that, if any of the folk beliefs about how best to conceive through ordinary intercourse in order to get a boy or girl turned out to actually work, one could hardly justify governments banning them from a couple's sex life. On the other hand, one might wonder if a child selected for gender would be burdened by an even greater degree of stereotypical expectations than we so often (despite our best intentions) impose on our children. It's complicated. The possibilities opened up by new medical techniques often are.

At any rate, once embryo selection beyond avoidance of genetic disease becomes an option, as it is already in some countries (the practice is not nationally regulated in the United States), the ethical and legal aspects are a minefield. When is it proper for governments to coerce people into, or prohibit them from, particular choices,

such as not selecting *for* a disability?* How can one balance individual freedoms and social consequences?

* * *

If there is going to be anything even vaguely resembling the popular
designer-baby fantasy, it will come first from embryo selection, not
genetic manipulation. "Almost everything you can accomplish by
gene editing, you can accomplish by embryo selection," says Greely.

But "designing" your baby by PGD looks currently unattractive,
as well as rather ineffectual. Egg harvesting is painful and invasive,
entailing a course of hormones to stimulate egg production and
then a very uncomfortable surgical procedure to extract them.
And it doesn't yield many eggs anyway: a typical IVF cycle might
produce perhaps 6 to 15, of which around half might become
fertilized *in vitro* to give embryos that look good enough for
potential implantation (generally only one or two are returned to
the uterus). There's not a lot of choice. And the success rate for
implanted embryos is still typically just 30 per cent or so at best.
The procedure is gruelling both physically and emotionally, and
expensive too, typically costing £3,000 to £5,000 per cycle in the
UK in 2018.

That's why no one currently chooses IVF unless they have to; it's
not by any stretch of the imagination a fun way to make a human
being. But some think that this may change. What if IVF was reliable, cheap and relatively painless? And what if it could allow you
to decide which baby you wanted?

That choice would look more meaningful if there were many
more options. Even if the outcomes were somewhat uncertain and

---

* According to one estimate in 2008, around one in 20 cases of PGD in IVF
clinics in the United States involves selection of embryos for conditions shared
by the biological parents that are generally classed as disabilities, such as dwarfism
and deafness.

probabilistic, might people not feel tempted by the prospect of having, say, a hundred embryos to select from, rather than simply accepting whatever fate provides? PGD on such a scale looks ever more possible as genetic screening becomes more affordable. Since the completion of the Human Genome Project, the cost of human whole-genome sequencing has plummeted. In 2009, it cost around $50,000; in 2017, it was more like $1500, which is why several private companies can now offer this service. In a few decades, it could cost just a few dollars per genome. Then it becomes feasible to think of conducting PGD on industrial-scale batches of embryos.

That's an uncomfortable conjunction of words, but an apt one if there were some way of obtaining and fertilizing many eggs at once. "The more eggs you can get, the more attractive PGD becomes," Greely says. One possibility is a once-for-all medical intervention that extracts a slice of a woman's ovary and freezes it for future ripening and harvesting of eggs. It sounds drastic but would not really be so much worse than current egg-extraction and embryo-implantation methods. And it could give access to thousands of eggs for future use.

But we saw earlier that there could one day be another option that requires no significant surgery at all: the manufacture of eggs *in vitro* by reprogramming somatic cells into gametes. Greely believes that a confluence of these two new technologies – making artificial gametes from the induced pluripotent stem cells of the prospective parents, and fast, cheap PGD of embryos to identify the "best" – could make what he calls Easy PGD the preferable option for procreating in the near future. He foresees "the end of sex" – not for recreational purposes but as a means of having babies. "The science for safe and effective Easy PGD is likely to exist some time in the next 20 to 40 years," he says – and he figures that on a similar timescale reproductive sex "will largely disappear, or at least decrease markedly."

So here are your hundred or thousand embryos, made from *in vitro*-generated eggs and each with a genetic profile obtained by PGD. What sort of child would you like, madam?

The catch is that these genetic menus will be phenomenally hard to assess and interpret. Some of the outcomes predicted from an embryo's genome will be definite, some likely, some just vaguely statistical. It might look something like this:

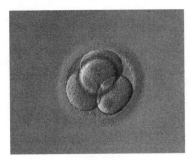

Embryo 78

- Male
- No serious early onset diseases, but a carrier for phenylketonuria [a metabolic malfunction that can cause behavioural and mental disorders]
- Higher than average risk of type 2 diabetes and colon cancer
- Lower than average risk of asthma and autism
- Dark eyes, light brown hair, male pattern baldness
- 40 per cent chance of top half in SAT tests

Let's say you have 200 of these to choose from. Beyond avoiding serious diseases, how on earth do you make that choice? How do you weigh up the pros and cons and balance them against analogous outcomes for another ten, or a hundred, embryos?

You can't. This is another example of the emptiness of the free-market mantra of choice.

For such reasons, some experts are sceptical of Greely's forecast of widespread use of PGD, even if it became "easy". "Where there is a serious problem, such as a deadly condition, or an existing obstacle, such as infertility, I would not be surprised to see people take advantage of technologies such as embryo selection," says bioethicist Alta Charo. "But we already have evidence that people do not flock to technologies when they can conceive without assistance." The low take-up rate for sperm banks offering "superior" sperm, she says, already shows that. For most women, "the emotional significance of reproduction outweighed any notion of 'optimization'." Charo feels that "our ability to love one another with all our imperfections and foibles outweighs any notion of 'improving' our children through genetics."

Not everyone is so confident. In the next 40 to 50 years, says Ronald Green, "we'll start seeing the use of gene editing and reproductive technologies for enhancement: blonde hair and blue eyes, improved athletic abilities, enhanced reading skills or numeracy, and so on." The allegedly HIV-resistant "CRISPR babies" made by He Jiankui are arguably already examples of this.

One factor that might help to spread something like Greely's Easy PGD is peer pressure. Goaded by the distorted agendas of current education policies, we are being encouraged to regard child-rearing as a competitive horse race in which we are made to feel irresponsible if we don't seize every advantage. If you'll invest your savings to move to a new house within the catchment area of a good school, or pay eye-watering private fees, why not a thousand or so more for a spot of genetic tailoring at the outset – even if the benefits are small at best?

Already the bullying rhetoric from gene-analysis companies has begun: "Genetic testing is a responsibility if you're having children,"

Anne Wojcicki, CEO of the US-based genome-sequencing company 23andMe, has asserted. Greely anticipates the kind of sloganeering that will seek to capitalize on these pressures: "You want the best for your kids, so why not have the best kids you can?"

Despite all the uncertainties, ambiguities and conflicting signals of a PGD analysis, it's not hard to see the temptations. Imagine that you are presented with just two embryos that have been screened by PGD, and you're told that one has a genetic profile matching the top 10 per cent for intelligence and the other the bottom 10 per cent. All else being relatively equal, which are you going to choose?

This is not a hypothetical scenario. It is already happening. In late 2018, an American company called Genomic Prediction announced that it would offer screening for IVF embryos to identify those whose genetic profiles suggest that a resulting child would have an IQ low enough to be classed as a mental disability. That in itself does not seem obviously any more objectionable than screening for disease or for Down's syndrome (which lowers IQ). But because it involves measuring the "polygenic" score (the summed effect of a great many genes) for intelligence, the test could just as readily be used to identify potentially high-IQ embryos. Genomic Prediction says it will not permit such uses – but even the company's co-founder, physicist Stephen Hsu, agrees that the demand will exist and that other companies in the unregulated United States are likely to satisfy it. Some experts objected that the science isn't (yet) up to supporting such claims – but do you think that will deter customers?

Beyond traits like intelligence, it may well be feasible to select embryos for all-round health. Greely thinks that a 10 to 20 per cent improvement in health via PGD is perfectly feasible. If so, isn't it reasonable to seek that for your child? Should a government be able to obstruct that choice?

But there are plenty of reasons, beyond safety considerations, to feel wary of permitting this kind of selection. The siren allure of perfection via Easy PGD could drive expectations to pathological extremes. What if the child genetically selected for athletic ability or artistic talent fails to deliver – as some inevitably will, given that such predictions are only probabilistic? You can hear the outrage of the parents: "Do you know what that treatment cost?!"

Some will argue that unequal availability of choice to different socioeconomic sectors of the population could seriously disturb social stability, leading to a "genetic divide" of the kind portrayed in the 1997 movie *Gattaca*. Greely warns that PGD to boost health, added to the comparable advantage that wealth already brings, could lead to a widening of the health gap between rich and poor, both within a society and between nations.

Yet others will say that prohibition of the reproductive choices that PGD (especially an "Easy" variant) offers is an infringement of rights. Might it not, indeed, be crazy or even immoral, if we have the means to improve the general intelligence of the population by making this choice available to all, to forego that chance?

Yes, it's complicated. I have my views, and I daresay you will too, but I think it is fair to say that the answers to these questions are not obvious. They demand that we find a balance between personal choice, liberty and morality versus the role of government: how states forbid, regulate or promote technologies. That of course is nothing more than the age-old question of democratic politics – except that now we are talking about how it might impinge on our capacity to shape human nature itself. "For better or worse, human beings will not forego the opportunity to take their evolution into their own hands," says Green. "Will that make our lives happier and better? I'm far from sure."

* * *

If you're seeking to conceive a child with intelligence or good looks, why bother with the lottery of sexual recombination of genomes in the hope of getting the right genes? Why not just make a copy of a smart or attractive person who you know already to have the requisite genetic make-up?

We saw earlier how cloning entails moving a fully formed set of chromosomes from one cell – which can be an adult somatic cell (in somatic-cell nuclear transfer or SCNT) – into an egg that has had its own nucleus removed. The egg is then stimulated in some way – chemical or electrical triggers will do the trick – to develop into an embryo under the guidance of the new chromosomes.

Cloning of animals did not start with Dolly the sheep. As we saw, Hans Spemann achieved it in the 1920s, first by dividing a salamander embryo in half with a fine noose, then using SCNT. Briggs and King did it for frogs in 1955. Sheep were first cloned that way in 1984; the big deal about Dolly was that the transplanted nucleus came from an adult somatic cell. In 2005, Woo Suk Hwang and his team in South Korea were the first to clone a dog, which they called Snuppy.* Hwang fell from grace soon after, when his claim to have cloned human embryos to generate stem cells was shown to be based on fraudulent data (see page 151).

We don't know if cloning is possible for reproductive purposes in humans. The only way to find out for sure is to try it, and that is banned in most countries;† a declaration by the United Nations in 2005 called on all states to prohibit it as "incompatible with

---

* Many wealthy pet owners, most famously Barbra Streisand, have since availed themselves of that facility. Some, however, have learnt the hard way that cloning does not guarantee identical copies. Entrepreneurs John Sperling and Lou Hawthorne were dismayed to find that the first cat cloned in a project in Texas, in which they invested in the hope of tapping a lucrative market, did not look like the donor. Inevitably dubbed CopyCat, it was black and white, lacking the orange of the tortoise-shell "original". This is because the fur colour of cats is controlled not genetically but epigenetically. It was a vivid illustration of how misguided is any notion that all that we prize in ourselves or other beings is genetically determined.
† Some countries, including the UK and Australia, permit non-reproductive *therapeutic* cloning to make human blastocyst embryos for research purposes

human dignity and the protection of human life". Yet a kind of human cloning was already conducted back in 1993, when scientists at the George Washington University Medical Center in Washington DC artificially divided human embryos produced by IVF: a kind of induced twinning of the sort that generates identical twins.* The cells grew into early-stage embryos but didn't reach the point where they could implant in a uterus. The work caused much controversy, because it wasn't clear if it had received proper ethical clearance.

Human cloning by somatic-cell nuclear transfer is another matter. The best reason we have so far to think it might be possible is a demonstration in 2017 that the technique works for other primates. Mu-ming Poo and his co-workers at the Institute of Neuroscience in Shanghai made two macaque monkey clones by SCNT, which they christened Hua Hua and Zhong Zhong. The donor cells came from macaque fetuses, not adult monkeys, but the researchers are confident that the latter will eventually work too. Primates have been particular difficult mammals to clone, for reasons not fully understood, and so the result represented a significant step towards human cloning. The work was conducted not as an end in itself but to create a strain of genetically identical monkeys that could aid research into the genetic roots of Alzheimer's disease.

Quite aside from the wider ethical issues, safety concerns would make it deeply unwise to attempt human cloning yet. Hua Hua and Zhong Zhong were the only live births from six pregnancies, resulting from the implantation of 79 cloned embryos into 21 surrogate mothers. Two baby macaques were in fact born from embryos cloned from adult cells, but both died – one from impaired body development, the other from respiratory failure.

– as Hwang claimed to have done.
* Actually this work used embryos with an excess of chromosomes, which meant there was no guarantee that the divided cells were genetically identical clones.

To construct a scenario where cloning seems a worthwhile option for human reproduction takes a lot of ingenuity anyway. One might perhaps imagine a heterosexual couple who want a biological child, but one of them will necessarily pass on some complex genetic disorder that can't be edited or screened out – so they opt to clone the "healthy" parent. But there's more than a hint of narcissism about such a hypothetical choice.

Of course, it's not hard to think up *bad* reasons for human cloning – most obviously, the vanity of imagining that one is somehow creating a "copy" of oneself and thereby prolonging one's life by proxy. That would not only be obnoxious but deluded. Even the idea that a clone will be a "perfect" reproduction of the person who supplies the DNA is mistaken. As we've seen, the "genetic programme" of a zygote is filtered and interpreted by the chance and contingency of development, with results that are not entirely predictable. Dolly the sheep was not the spitting image of the ewe from which she was cloned in 1997, and four ewes cloned by the same team at the Roslin Institute in Scotland two years previously from embryonic donor cells were, according to the researchers, "very different in size and temperament". A clone of Einstein would by no means be a genius of equal measure.

It will take some effort to dislodge the naïve genetic determinism that surrounds talk of cloning, though – and not just because of absurd fantasies like the cloning of Hitler in Ira Levin's 1976 novel *The Boys from Brazil*. Scientists will have to mind their language – to desist from calling the genome a "blueprint that makes us who we are", or (as Francis Collins, the head of the US National Institutes of Health recently said apropos the CRISPR editing of human embryos) "the very essence of humanity". That kind of talk is now dangerously misleading.

Claims from mavericks and cults notwithstanding, no human being has been cloned to date. But my feeling is that this will happen eventually. It's not a prospect I welcome, because (unlike IVF) there does not seem to be any sound justification for it, motivated solely by the mitigation of suffering and the welfare of a person created this way. All the same, if it comes to pass then we should anticipate another "Louise Brown moment", where we struggle to reconcile a sense of unease at an unfamiliar process of making humans with what is likely to be the evidence that it makes people just like us.

Ronald Green may well be right in principle, if not in terms of timescale, to have suggested in 2001 that within one or two decades "around the world a modest number of children (several hundred to several thousand) will be born each year as a result of somatic cell nuclear transfer cloning." It is quite possible that, as Green says, within several decades "cloning will have come to be looked on as just one more available technique of assisted reproduction among the many in use." I would rather see that being done in an open and regulated manner, with proper safeguards, than in remote locations by mavericks and profit-hungry companies who care little for the motives or perhaps even the welfare of their clients. Just because we might be wary of human cloning, that does not mean there is the slightest reason to be wary of a human made this way.

* * *

The possibilities for growing and shaping human beings provided by our new technologies for manipulating cells might seem dramatic, even alarming, but they are rather conservative compared to what some of scientists in the early days of the field foresaw.

John Desmond Bernal was one of that remarkable generation of biologists and biochemists from between the wars – it includes J. B. S. Haldane, Joseph Needham, Julian Huxley and Conrad Hal Waddington – who laid the foundations of both developmental and molecular biology while at the same time weaving those advances into a committed vision of the social roles of science. Bernal's extended essay *The World, the Flesh and the Devil* (1929) was a response to Haldane's speculations on the opportunities afforded by the discipline we today call biotechnology, which were themselves informed by research on tissue culture like that at the Strangeways laboratory in Cambridge.

If Haldane's forecasts of the future of procreation in *Daedalus, or Science & the Future* (1924) displayed a readiness to speculate and extrapolate way beyond what many scientists would be willing to risk today, that was nothing compared to Bernal's thoughts on where biotechnology might take us. We might, he said, eventually get rid of "the useless parts of the body" and replace them with mechanical devices: artificial limbs and sensory devices that do a much better job. In the end, this cyborg existence would mutate into the kind of "brain in a vat" that I discuss in the next Interlude, hooked up to a distributed system of engineered devices in lieu of a body:

> Instead of the present body structure we should have the whole framework of some very rigid material, probably not metal but one of the new fibrous substances. In shape it will be rather a short cylinder. Inside the cylinder, and supported very carefully to prevent shock, is the brain with its nerve connections, immersed in a liquid of the nature of cerebro-spinal fluid, kept circulating over it at uniform temperature. The brain and nerve cells are kept supplied with fresh oxygenated blood through the arteries and veins which connect outside the cylinder to the artificial heart-lung digestive system – an elaborate, automatic contrivance.

Bernal's ideas are clearly indebted to the science fiction of his era. Two years before his book was published, *Amazing Stories* carried a tale by the pseudonymous Francis Flagg called "The Machine

Man of Ardathia", in which a modern-day American encounters a visitor from the future: a fetal humanoid encased in glass and wired up to machinery. Olaf Stapledon's science-fiction classic *Last and First Men*, published in the same year as Bernal's essay, described how humans in the future would use Haldane-style artificial conception along with Aldous Huxley-style biological manipulation to make entities that are more or less immense brains with tiny residual bodies tangling like an appendage from the under-surface, kept alive by pumps perfusing oxygenated blood. Eventually these man–machine hybrids – the Fourth Men, in Stapledon's elaborate post-evolutionary genealogy of humankind – evolve into enormous brains housed in towers 40 feet across: disembodied intelligences that enslave the earlier anthropomorphic humans (Third Men) and colonize the planet from the deepest seabed to the atmosphere.* These fantasies are obviously influenced too by the work of Alexis Carrel, who himself envisaged a kind of distributed body in which the organs were kept in separate vessels but connected by a vasculature of tubing.

Bernal's speculations are now regarded as a part of the intellectual heritage of the movement known as transhumanism, which seeks to use technologies to extend the possibilities of the human body in radical ways. The definition offered in 1990 by Max More, CEO of the cryonics company Alcor Life Extension Foundation, remains as good as any; it says that transhumanism refers to:

> Philosophies of life . . . that seek the continuation and acceleration
> of the evolution of intelligent life beyond its currently human form

---

* Stapledon's book – I hesitate to call it a novel, because it is a dense but plotless chronicle of humankind's far future – is a remarkable flight of the imagination. In an anticipation of the technologies discussed in this book that some might consider chillingly prescient, Stapledon explains that the super-brains of the Fourth Men are produced by "manipulation of the hereditary factors in germ

and the human limitations by means of science and technology, guided by life-promoting principles and values.

Despite what is sometimes asserted, transhumanism is not about "perfecting the human body". Its proponents generally recognize no platonic perfect form but believe that improvements can continue indefinitely (which after all is evolution's position too). Much of the transhumanist programme so far has focused on the extension of cognitive and sensory capabilities using medical and information technologies, drugs, and human–machine interfaces. The plasticity of human flesh itself has become a largely unforeseen addition to the transhumanists' utopian arsenal.

Of course, much rests on that issue of More's "life-promoting principles and values", since there is no agreement on what these comprise, nor any fixed philosophical or ethical calculus for resolving the matter. Transhumanists usually tend towards liberal beliefs, grading into full-blown libertarianism – and they may find themselves challenged by the stark fact that our narratives of efforts in this direction veer almost uniformly towards the dystopian.

Transhumanists are sometimes accused of finding the human body repulsive. They deny it, and we should do them the courtesy of believing them – but the body tends to feature in their plans as flawed at best, and perhaps as an unnecessary encumbrance. Even the brain itself is commonly viewed purely as an information-processor whose job can be done just as well by a computer, a contention that is still controversial within neuroscience, as we'll see.

---

cells (cultivated in the laboratory), manipulation of the fertilized ovum (cultivated also in the laboratory), and manipulations of the growing body." Stapledon suggests, however, that the Fourth Men eventually come to realize that their life of pure intellect is constrained and devalued precisely by the lack of a body. For that reason, they engineer their successor from the remnants of the embodied Third Men: "he was to be a normal human organism, with all the bodily functions of the natural type; but he was to be perfected through and through."

A contempt for the flesh is certainly apparent in another book commonly cited as heralding the transhumanist project: *Man into Superman* (1972) by American cryonics visionary Robert Ettinger. If Bernal's proto-transhumanism merged seamlessly with the imagery of *Amazing Stories*, so too does Ettinger's book reflect the spirit of its age, for it is a lurid acid trip, exuding the kind of self-assurance characteristic of the most extreme and unhinged scientific fantasies. His delirious speculations about the future sex of bioengineered bodies could have come from the pages of a J. P. Donleavy novel:

> The sexual superwoman may be riddled with cleverly designed orifices of various kinds, something like a wriggly Swiss cheese, but shapelier and more fragrant; and her supermate may sprout assorted protuberances, so that they intertwine and roll all over each other in a million permutations of The Act, tireless as hydraulic pumps . . . A perpetual grapple, no holes barred, could produce a continuous state of multiple orgasm.

While the language of today's transhumanism is rather more sober, it sometimes conveys a similar frustration that our all-too-human flesh inhibits our sexual potential. Biotech entrepreneur Martine Rothblatt looks forward to the creation of "digital people", the *personae creatus*, who are liberated from the conventions of sexuality and gender that bodies impose. "Once that is done," she writes, "sexual identity will be liberated not only from genitals, but from flesh itself. Consciousness will be as free to flow beyond the confines of one's flesh as gender is free to flow beyond the confines of one's flesh genital." There would be an end to the straitjacket of sexual dualism.

That's as explicit an example as you could imagine of transhumanism as displaced wish fulfilment. What it seems to represent for Rothblatt is an embodiment of the desire for society to be less rigid and definitive in the way it regards sexuality and gender. It's a valid aspiration, but there are already sound biological and cultural reasons why this should apply equally to our fleshy selves, without any need to invent a digital diversity of sexual categories. Here, as elsewhere, transhumanism seems to function more as an imaginarium for constructing the utopias we might (with good reason) wish to see realized in the world as it already is.

Much the same can be said for the transhumanist obsession with immortality. Ettinger's position that "humanity itself is a disease, of which we must now proceed to cure ourselves" is, at least so far as our mortality is concerned, one espoused by plenty of transhumanists today. And his spirit surely hovers over the transhumanist view that an acceptance of death is a kind of craven and mystical nihilism:

> Those who are willing to settle for mortality and humanity just do not understand their predicament or their opportunity, how lowly they are and how exalted they may become.

Some transhumanists seek immortality through biomedical avoidance or reversal of the ageing process, via diet, drugs, lifestyle choices and surgical intervention. Others hope that their consciousness might be downloaded from the brain into a hard drive. Others arrange for their bodies to be frozen after death by companies like Max More's, in the hope of resurrection once technology has acquired the capabilities. (Ettinger himself was cryopreserved, as were his first and second wife.) But we can now see that the ancient allure of immortality has long complicated any deep engagement with, and confrontation of, our fleshy selves. Senescence, obsolescence and morbidity are built into our cells – but their versatility seems also to offer a promise of rejuvenation, whether that is by reversing the process of differentiation and returning to an embryo-like state, or reawakening the process of cell proliferation, or transferring our DNA (our essence, our soul, so we are told) to some fresh new vessel in the process of cloning.

When we consider that life as a process has persisted resolutely for around four billion years, it seems odd that flesh, through the lens of transhumanism, is seen as weak, fragile and ephemeral, and that the self is thought to be housed more robustly in inorganic material: in Bernal's glass and steel, Stapledon's "ferro-concrete", and now in silicon circuitry.* Of course, the longevity of life in the evolutionary sense is scant consolation to transhumanists, who want their own *individual* lives to be sustained through the ages. But as the final chapter explains, it is by no means clear that there *is* any concrete, static and boundaried individuality of being in the biological sense. What they wish to preserve is not a thing reducible to some instantaneous configuration of bytes, but a process that is intrinsically embodied, dynamic, ephemeral, contingent, and tied to its environment. That is what being alive means. You might as well dream of storing the flow of a river.

And so transhumanism seems often to oscillate between a peevish impatience with the human body (and with those of us who are reconciled to it) and a slightly patronizing assurance that, while nature has made a noble and creditable start, we can do better. Max More's "Letter to Mother Nature" gently chides her: "What you have made us is glorious, yet deeply flawed. You seem to have lost interest in our further evolution some 100,000 years ago . . . We have decided that it is time to amend the human constitution." This will look to some like sheer Frankensteinian hubris, and likely to end no better. But we have been influencing our own evolution for

---

* I contemplate the old floppy disks at the bottom of my drawer, crammed with data that are useless, inaccessible and no doubt degrading, and I wonder how long-lived a hypothetical "virtual self" would truly be if hosted digitally.

centuries, and we already have capabilities for making that haphazard process a matter of design. One danger of transhumanism is not that it poses hubristic questions and challenges – today's biotechnologies do that already – but that it is so readily appropriated by false prophets and technological fantasists in pursuit of their own obsessions, or in flight from their terrors.

Sure, it's easy to knock transhumanism, not least because the future outlined by its advocates so often seems like a dour form of solipsistic hedonism in which intellectual excellence has been sapped of all joy and humour. But at the very least, it outlines a thought experiment worthy of serious ethical reflection. We already put a great deal of effort into seeking what we consider to be the "good life": extending the period of good bodily and mental health, cultivating meaningful relationships, alleviating suffering in others, respecting individual autonomy and rights, deepening out intellectual and emotional engagement with the world. If the technologies of medicine and information can afford new opportunities towards these goals, why would it be anything but ethically wise and responsible to take them?

What is more, it is hard to argue against the transhumanist principle of "morphological freedom" – as a Transhumanist Manifesto written in 1998 by several champions of the cause puts it, "the right to enhance one's body, cognition and emotions." As with reproductive genome editing, such freedoms raise difficult questions of social equality and access, which some at the libertarian end of the transhumanist spectrum seem disinclined to pursue. But the principle of redesigning the body to extend its capabilities is nothing more than we have practised for centuries, at least since the development of prostheses and aids for vision and hearing. While medical technologies of this kind developed for remedial purposes, such as artificial limbs that respond to nerve impulses or eye-tracking screens, can supply dramatic results, so far most efforts aimed at human enhancement have tended towards the trivial or gimmicky, such as implanted radiofrequency devices that can activate external security circuits. But technologies of cell transformation might, as we have seen, soon render possible some remarkable morphological changes of our fleshy body, and transhumanism might motivate useful and even essential debate about what is possible and desirable.

What seems to me the major impediment to most visions of transhumanism is not, then, the nature of the goals but the lack of technical and biological realism attached to the means of attaining them. Many of the imagined futures of today's transhumanists would not look out of place alongside the wild and wonderful fantasies that graced *Amazing Stories* in the 1930s. They too often rest on simplistic and optimistic extrapolations, or on purely fantastical technologies. There is commonly a bathetic gulf between the possibilities that transhumanists imagine and what is actually being done (and what limitations are known to exist) in areas such as neuroscience, cognition, information technology, biomedical engineering and nanotechnology, let alone what role socioeconomic factors will play in shaping how they evolve. As a result, even some relatively recent writings in transhumanism can look quaintly misguided a decade later.

This is not to say that the aims and forecasts of transhumanism are worthless. But their worth often lies in a different direction to what the advocates intend: as a looking glass for our hopes, dreams and fears. In that respect the field is surely no different from – and no better or worse than – any attempt to glimpse the technological landscape ahead.

PHILOSOPHY OF THE LONELY MIND

CAN A BRAIN EXIST IN A DISH?

As my mini-brain in a dish took shape, I found myself recalling the fledgling novel that, many years ago and quite wisely, I buried in a drawer and forgot. It contained a scene satirizing the Grand Guignol of *Amazing Stories*, in which a crazed doctor is building a gigantic brain, sustained by perfusion of blood, in the basement of a hospital by merging the organs he removes from hapless patients. Ultimately, he plans to house it in an immense, grotesquely designed synthetic head with multiple pairs of eyes, no nose, that sort of thing.

Fear not: this hideous progeny is not about to re-emerge from my bottom drawer. But while Selina Wray does not exactly fit the persona of my deranged Dr Zoback, what she and Chris Lovejoy (oh yes, there was a lunatic henchman too . . .) had created in the incubators of UCL was in some ways far stranger, and certainly more wonderful, than anything my juvenile mind conjured up. All I was doing back then was toying with an old and hackneyed trope. We find it, for example, in a story by M. M. Hasta called "The Talking Brain" from a 1926 issue of *Amazing Stories*, which was directly inspired by Alexis

Carrel's work. "If a heart could be kept beating in a bottle for years at a time," asks Hasta's obsessed scientist Professor Murtha, "why should not a brain be kept thinking in a bottle forever?"

The reader can, of course, immediately think of several good reasons why not. But such considerations don't stop Murtha from removing the brain of a student who was severely injured in a car crash and placing it in a wax head. Murtha wires up the perfused brain so that it can communicate using Morse code, and . . . well, you can guess, can't you? It says:

> This place is more terrible than you can know. Set me free. Kill me or let him [Murtha] kill me. Now. Now. Now. Now. Now. Now. Now.

I don't know if Roald Dahl ever read Hasta's story, but he was of course perfectly capable of thinking up that fiendish scenario on his own. It appears in Dahl's short story "William and Mary", published in 1959 and later used for an episode of the TV series *Roald Dahl's Tales of the Unexpected*. William is a philosopher who knows he will shortly die of cancer. He is approached by a doctor who offers to preserve his brain after death, hooked up to an artificial heart and to one eye, like a nightmare version of Carrel's experiments. William agrees to this proposal, and all goes ahead, William's wife Mary having agreed to look after the disembodied assembly of organs her husband has become. But Mary has plans. After a lifetime of boorish domination by her husband, who forbade her to smoke or to get a television, she now intends to do those very things with the helpless brain looking on. The lone eye stares at her, somehow registering fury; she calmly blows smoke in it. "From now on, my pet, you're going to do just exactly what Mary tells you," she purrs. "Do you understand that?"

But there are more uses for a "brain in a jar" than making the skin crawl. The image has tantalized philosophers for a long time, who weave it into scenarios through which to examine belief and scepticism. They used to regard this as a mere (forgive me) thought experiment. But thought experiments have a habit of turning into real ones.

* * *

Nothing about the brain quite makes sense to our intuition. Here is a surgeon holding in her hands a human brain freshly removed from its cranium, and it seems like so much offal from a butcher's slab, a labyrinthine blancmange with bloodied crannies. (Not until the formalin preservative gets to work does the tissue attain the rubbery consistency of the specimen jar.) The mere pressure of a finger will dent the juddering tissue.

And yet in that bland mass was once a universe and a lifetime. Everything that person knew and felt and experienced – the sound of breakers on a tropical beach, the taste of roast chestnuts, the pain of a mother's death – was locked within those membranes, coded in electrical patterns travelling within a substance about as different from our typical image of an information-processing device as it could be. You can't tell me there is not something mysterious about this.

The brain conjures up worlds with the skill of a master illusionist. I can tell myself that the keyboard I see before me, the street outside the window, are the objective world – but their colours, their distinctness and perspective, every aspect of what I perceive them to be, are constructed inside my head. I could never convince myself to fully accept that truth. I can't even talk about it coherently; here I am creating some little homunculus that, like the *Beano*'s Numskulls, lives in my brain and is fed sensory data by the blobs of grey matter. The brain is the ultimate philosophical puzzle. This is the seat of person-hood, yet it's not clear who is to be found there, or where they are.

We can't help imagining that within the soft clefts of the human brain somehow resides the person themselves – or at least clues to what made them who they were. Albert Einstein's brain, removed by pathologist Thomas Stoltz Harvey after the physicist's death in Princeton in 1955, was cut into slices and preserved. Harvey himself hoarded some of those fragments. Others have now found their way into museums, where they have become macabre emblems of genius akin to the alleged somatic relics of saints. Claims abound, some coming from serious neuropathologists, about why Einstein's brain was anatomically "special". But the truth is that everyone's brain is likely to show some deviations from the norm, and attributing specific abilities to this or that variation in shape, size and cellular structure of brain tissue can be a hazardous enterprise. We don't know what made Einstein intellectually exceptional, but it's not obvious that answers can be found by rummaging around in his preserved grey matter.

But even if you can't make deductions about character or genius from the general topology of the brain, still there is some truth in the notion that our brain makes us what we are. The brain gets hard-wired as we develop, guided (but not prescribed) by our genes, and from someone's genetic profile you can make meaningful predictions about likely character traits. But as we saw earlier, these are probabilities, not destinies. Brain development is contingent on the sensitivities of this complex genetic machinery to random noise accidents of growth – you can't tell exactly how it will turn out.

The brain is a modular organ: certain behaviours can be linked to physical features of different brain regions. That's why damage to specific areas can have very particular consequences for cognitive function. There are clear, albeit minor, anatomical differences between male and female brains, although what that might imply for putative behavioural differences is still debated. And atrophy of specific brain regions can result in highly specific symptoms.

Dementia, for example, is not just about generalized loss of memory; there are many types, with distinct symptoms linked to the parts of the brain that are affected. Primary progressive aphasia (PPA), produced by deterioration of the frontotemporal lobes, affects aspects of semantic processing: how we label concepts, or our ability to retrieve those linguistic tags. For some people living with PPA this manifests as a struggle to connect words to sound production. But if the neurodegeneration of PPA happens a little further back in the part of the brain (the temporal lobes) where language is processed, what gets impaired is not the forming of words but accessing their semantic content.

In contrast, posterior cortical atrophy (PCA), a variant of Alzheimer's disease, commonly affects spatial awareness and can lead to disorientation, visual illusions and coordination problems. The experience can be akin to that moment when we're unable to parse what we see: is that a face, or just the folds in a piece of fabric? Is that object near or far? One man with PCA who I spoke to during the Created Out of Mind project that spawned my "brain in a dish" told me how the piano keyboard he was playing seemed on one occasion to rise up a couple of feet higher. Such experiences might be best viewed not as a "distortion of reality" but a reminder of how much of a mental construct all perception is.

The brain is shaped by experience. Musicians who began their training at a young age show enlargement of the region called the corpus callosum, which connects and integrates the processing of the separate hemispheres. The part of the cortex used to process musical pitch also shows boosted development in musicians. Brain scans of London taxi drivers show that they have enhanced development of the rear of the hippocampus, a region associated with memory and navigation. This enlargement happens in proportion to the amount of training that the drivers have had in learning the city's street layout – so it seems to be a genuine response to, and not a cause of, their skill at finding their way around. In so many ways, our brains, like our bodies, are not a given fleshy mechanism by means of which we navigate the world, but an adaptive and responsive record of our personal histories.

*　*　*

Who can forget Frankenstein's crazed assistant Fritz giving him the "abnormal" brain of a criminal in James Whale's 1931 movie of Mary Shelley's story?* This, we are supposed to infer, accounts for the monster's murderous tendencies, totally undermining Shelley's message that the creature is turned that way by Frankenstein's failure to give his creation care and love.

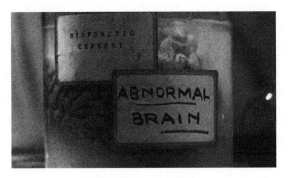

No, not that one Fritz: a still from Frankenstein *(1931).*

The preserved brain in a jar was no Hollywood invention, though. The organs were often stored this way for medical research, to be analysed for visible features that might be linked to behaviour patterns. Anatomical dissection seems now a rather crude approach

---

* Who, for that matter, can forget Marty Feldman admitting to Gene Wilder's doctor in *Young Frankenstein* (1974) that the bottled brain he has purloined belonged to "Abby Normal"?

to understanding such a phenomenally complex organ, but it was once the only way to find out about associations of form and function in the body.

There are exceptions to that respectable tradition of tissue preservation far more grotesque and horrific than anything you find in Gothic horror films. In the 1970s, shelves of jars containing the preserved brains of hundreds of children were found to have been kept in the basement of the Otto-Wagner Hospital in Vienna, where attendants had unquestioningly been topping up the preserving fluid for decades. These brains, it transpired, had been removed from children kept in a special ward and murdered as "mental defectives" by command of the Nazi doctor Heinrich Gross. Gross apparently intended to study the anatomical causes of such "defects". The remains of the child victims were buried in a ceremony in 2002.

Some people want their brain to end up in a jar by choice – not for the benefit of medical research, but because they figure they might one day need it again. Or at least what's allegedly stored *in* it. Brain freezing is big business: many hundreds of people have paid for their bodies (or just their heads – the budget option) to be cryogenically preserved after death in the hope that science will one day enable the brain to be revived and the person in effect brought back to life. You won't necessarily need or want your original body, especially if you died from some fatal accident or illness. According to Alcor Life Extension Foundation in Scottsville, Arizona, one of the leading companies to offer this service, "When carried out under favorable conditions, both current evidence and current theory support the conclusion that cryonics has a reasonable chance of working." (It's not clear quite what this "current evidence and theory" are.) The company adds that:

Survival of your memories and personality depend on the extent of survival of brain structures that store your memories and other identity-critical information . . . It cannot be reliably known with present scientific knowledge how a given degree of preservation would translate to a given degree of memory retention after extensive repair, but sophisticated future recovery techniques using advanced technology might allow for memory recovery even after damage that today might make many think there was little room for hope.

Well, it's your choice – at a minimum cost of $80,000 per capita (never has that Latin term been more literal). Some critics accuse human-cryopreservation companies of profiting from false hopes peddled to vulnerable people. Experts point out that today's cryogenic techniques inevitably cause damage to tissues, and that thawing would damage them more, so there is no prospect that a frozen brain could be revived.

Respectable cryonics companies like Alcor have tried to distance themselves from the more far-fetched claims of the brain-freezing immortalists, but nonetheless such technologies are vaunted as offering a glimmer of hope that death can one day be cheated. The practice of post-mortem brain-freezing became especially controversial when, in 2016, the British High Court ruled to respect the expressed wish of a 14-year-old girl who died from cancer to have her body cryogenically frozen by an American company so that she might one day have a chance of being "woken up".*

---

* It's a tragic tale for sure, but one can't discuss the legalities of an individual being perpetuated in their isolated and preserved brain without mentioning this court exchange allegedly reported in the Massachusetts Bar Association Lawyers' Journal:
   Lawyer: Doctor, before you performed the autopsy, did you check for a pulse?
   Doctor: No.
   L: Did you check for blood pressure?
   D: No.
   L: Did you check for breathing?

Immortalists might take some comfort, however ghoulish, from controversial experiments by researchers at Yale University in 2018 in which decapitated pigs' heads taken from an abattoir shortly after slaughter were reportedly kept alive in some sense for 36 hours by perfusion with an oxygenated fluid. "Alive" here meant merely that some of the cells showed signs of surviving during the perfusion; the heads did not – mercifully, but unsurprisingly – regain consciousness.

Sustaining cell activity in a head recently removed from the body is a long way from reviving memories in a head deep-frozen for many years. But some advocates think it is not a vain hope that we might one day have the technical means to do that. "If you can bridge the gap (it's only a few decades)," writes computer scientist Ralph Merkle, "then you've got it made. All you have to do is freeze your system state if a crash occurs and wait for the crash recovery technology to be developed . . . you can be suspended until you can be uploaded."

Wait – a crash? Uploaded? You can see where this is going: towards the idea that the brain is just a kind of computer and can be described using the terminology of the laptop.

To Merkle it's just a matter of physics. Your brain is made of material, and so is governed by the laws of physics; those laws can be simulated on a computer, and therefore your brain can be too. You don't even need squishy, fragile grey matter – it's all just a question of bits and bytes. The network of neural connections in the brain is astronomically complex, but all the same we can put

---

D: No.

L: So, then it is possible that the patient was alive when you began the autopsy?

D: No.

L: How can you be so sure, Doctor?

D: Because his brain was sitting on my desk in a jar.

L: But could the patient have still been alive, nevertheless?

D: It is possible that he could have been alive and practising law somewhere.

an upper limit on how many bits should be needed to encode it. Merkle calculates that uploading a brain will need a computer memory of about a million trillion bits, performing around ten thousand trillion logic operations a second. That's at least imaginable with the current rate of technological advance. According to this transhumanist vision, we will soon be able to live on, inside computer hardware: the brain in a jar becomes the brain on a chip.

Whether a cryogenically frozen brain can preserve all the data responsible for your experiences and mental states during life is another matter (even setting aside the problem of freeze-induced damage). How signalling neurons conspire to create consciousness is one of the big scientific unknowns, and this number-crunching of bit counts is a little like describing the global economy simply as the net sum of its monetary value, or by counting up the trading agents involved. Some cognitive scientists dispute the idea that the human brain is merely a sophisticated kind of computer that can be mapped onto any other substrate if it has enough circuitry. They say that consciousness might be a property of very specific kinds of processing networks, and not necessarily compatible with silicon or digital hardware. To neuroscientist Christof Koch of the Allen Institute for Brain Science in Seattle, the idea that conscious thought is just a kind of computation is "the dominant myth of our age".

What such heady visions of brain downloads also ignore is that the brain is not the hardware of the self, but an organ of the body. Several experts in both artificial intelligence and cognitive science now argue that embodiment is central to experience and brain function. At the physiological level, the brain doesn't just control the rest of the body but engages in many-channelled discourse with its sensory experience, for example via hormones in the bloodstream. And embodiment is central to thought itself, according to artificial-intelligence expert Murray Shanahan of Imperial College London. Shanahan says that cognition is largely about figuring out the possible consequences of physical actions we might make in the world: a process of "inner rehearsal" of imaginary future scenarios. The brain doesn't just conduct such imagery in the abstract: during cognition, we see activation of the very same brain regions, such as those controlling motor function, that we would use if we were to carry out the imagined actions for real.

That fits with the suggestion of cognitive scientists Anil Seth and Manos Tsakiris that the function of the brain is not to compute sensory information in some abstract sense, but to use it to construct a model of the world that is consistent with our bodily experience of it. In other words, we don't create some bottom-up representation from basic sensory data, but start top-down with our sense of "being a body" and figure out what kind of world is consistent with that. We infer (subconsciously) that objects have an opposite side we can't see, not because we have learnt that "objects have hidden sides" but because we deduce that we'll see it if we walk around in the physical space.

It's for this reason, say Seth and Tsakiris, that we *don't* perceive ourselves as a Numskull-like homunculus looking out of a body-machine from the window of the eyes. Instead, we take the body for granted as an aspect of our mental world-map: it is woven tightly into our sense of self. Any aspect of our physiology that affects the integrity and function of our body – our immune system, say, or the microbes in our guts – then contributes to this sense of selfhood. In short, there can be no self without the somatic component: without a body.

In this view, the "brain in a jar" is not a feasible avatar of the entire human. One could argue that the transhumanist idea of a brain-on-a-chip could be coupled to a robotic body that allows such physical interaction with the surroundings, or even to just a simulation of a virtual environment. But the somatic view of self-hood raises questions about whether there is any purely mental "essence of you" that can be bottled and downloaded in the first place, independent of its embodiment in an environment.

The more interesting and relevant version of transhumanist self-realization, then, considers what happens to mind when body is extended and altered, given new senses or new capabilities by new interfaces between human and machine. Donna Haraway's classic 1985 essay "The Cyborg Manifesto" argued that we are at this point already. The boundaries between natural body and artificial mechanism (as well as between human and animal), she said, became increasingly blurred throughout the twentieth century. Like many transhumanists, she welcomed this development as a liberation from the "antagonistic dualisms", for example in gender and race, that society constructs. For some, this notion of the human increasingly embedded within the machine is unsettling: science-fiction writer Bruce Sterling, who pioneered the cyberpunk genre, imagines a wretched being who is as trapped and constrained as the vat-minds of *The Matrix*: "old, weak, vulnerable, pitifully limited, possibly senile". But it is precisely because we are "profoundly embodied agents", not "minds in bodies", says philosopher Andy Clark, that by extending embodiment by whatever means available – computational, genetic, chemical, mechanical – we will necessarily extend the self. We are, says Clark, "agents whose boundaries and components are forever negotiable, and for whom body, thinking, and sensing are woven flexibly from the whole cloth of situated, intentional action." The technologies of cell transformation have added a new and powerful axis of negotiation – and as we have seen, they recognize no distinction between the brain and body. All are contingent tissues, ripe for change. Clark's comment serves as a reminder that as we alter the body, we should expect the mind to change too.

* * *

The brain in a vat is a familiar image to philosophers, who have long used it as a vehicle for exploring the epistemological question of how we can develop reliable notions of truth about the world.

How can you be sure, they ask, that you're *not* just a brain in a vat, being fed stimuli that merely simulate a world? How can you know that all your beliefs about the world are not falsehoods shaped by illusory data?

The question goes back to the sceptical enquiry of René Descartes in the seventeenth century. Descartes argued that, since we can only develop a picture of the objective world from sensory impressions, we might just be being systematically deceived by an evil demon. In his 1641 *Meditations on First Philosophy*, he imagined that a being of "utmost power and cunning has employed all his energies in order to deceive me." In that event, he said,

> I shall think that the sky, the air, the earth, colours, shapes, sounds and all external things are merely the delusions of dreams which he has devised to ensnare my judgement. I shall consider myself as not having hands or eyes, or flesh, or blood or senses, but as falsely believing that I have all these things.

For Descartes this was not so much a disconcerting possibility that he had to argue his way out of, but a reason to doubt his convictions. It was through such scepticism that he devised his famous, reductive formula for the one thing of which he could be sure: I think, therefore I am.

Demons often did the job of constructing counterfactuals and conundrums for thinkers from the Enlightenment onwards. Pierre-Simon Laplace had one (which foretold the future of the universe from a complete knowledge of its present state), and so did James Clerk Maxwell (which undermined the second law of thermodynamics). In the twentieth century, scientific scenarios were deemed a more proper thinking tool than supernatural ones. Thus in 1973, the American philosopher Gilbert Harman recast Descartes's problem in the form of a medical scenario:

It might be suggested that you have not the slightest reason to believe that you are in the surroundings you suppose you are in . . . various hypotheses could explain how things look and feel. You might be sound asleep and dreaming or a playful brain surgeon might be giving you these experiences by stimulating your cortex in a special way. You might really be stretched out on a table in his laboratory with wires running into your head from a large computer. Perhaps you have always been on that table. Perhaps you are quite a different person from what you seem.

Here the legacy of the brain in a vat becomes apparent. For what is this but the underlying premise of the *Matrix* movies? By the time the Wachowski brothers (now sisters) began riffing on it, "brain in a vat" scepticism about what we take to be reality had a thorough philosophical pedigree. The most celebrated critic of the idea was American philosopher Hilary Putnam, who argued in 1981 that the proposition that you might be a mere brain in a vat fails because, in effect, it dissolves into contradiction. Words used by a brain in a vat can't be meaningfully applied to real objects outside of the brain's experience, Putnam said. Even if there are actual trees in the world containing the vat that are simulated for the brain, the concept "tree" can't be said to refer to them from the brain's point of view. Philosopher Lance Hickey encapsulates Putnam's argument with what I suspect is an unintended whiff of absurdist humour:

1. Assume we are brains in a vat.
2. If we are brains in a vat, then "brain" does not refer to brain, and "vat" does not refer to vat.
3. If "brain in a vat" does not refer to brains in a vat, then "we are brains in a vat" is false.
4. Thus, if we are brains in a vat, then the sentence "We are brains in a vat" is false.

Take this slowly. Or alternatively, take Anthony Brueckner's neat précis of the argument in the title of his 1992 paper on the issue: "If I Am a Brain in a Vat, Then I Am Not a Brain in a Vat."

Not everyone is persuaded by Putnam's rather elusive reasoning. Philosopher Thomas Nagel adds to the impression that philosophers seem here to be attempting to escape, Houdini-like, from the sealed glass jar of their own minds with the somersaulting logic of his riposte to Putnam:

> If I accept the argument, I must conclude that a brain in a vat can't think truly that it is a brain in a vat, even though others can think this about it. What follows? Only that I cannot express my skepticism by saying "Perhaps I am a brain in a vat." Instead I must say "Perhaps I can't even think the truth about what I am, because I lack the necessary concepts and my circumstances make it impossible for me to acquire them!" If this doesn't qualify as skepticism, I don't know what does.

No wonder Neo just decided to shoot his way out of the problem. Feel free to decide for yourself how convincing you find these arguments, for there is still no real consensus about whether we must just accept the world as we find it or can be sure we aren't the dupes of Descartes's demon. Technological advances now offer new variants of the same basic conundrum. Might we be virtual agents in the computer simulations of some advanced intelligence: one of countless simulated worlds, perhaps, created as an experiment in social science? According to one argument, it is overwhelmingly likely that we are just that, for if such simulations are possible (and some people think we are not so far off from being able to build a primitive degree of cognition into the avatars of our own virtual worlds) then there will be many more of them than the unique "real world".

Some philosophers and scientists have even considered how we might spot fingerprints of a simulated character in what we take to be the physical world, like the glitches that feature in *The Matrix*. But I can't help wondering what, to a simulated avatar, any notion of a "world outside" could mean – as if once again there exists within the experiential self some homunculus that can climb outside it. I wonder too at the readiness to make these supposed super-advanced intelligences merely smarter versions of ourselves: cosmic doyens of computer gaming. We have lost none of Descartes's solipsism; perhaps that's the human condition.

The "brain in a vat" might sound like one of those *reductio ad absurdum* scenarios for which philosophers enjoy notoriety, but some think it is already a reality. Anthropologist Hélène Mialet used precisely that expression to describe British physicist Stephen Hawking on the occasion of his seventy-first birthday in 2013. Hawking was famously confined to a wheelchair for decades by motor neurone (or Lou Gehrig's) disease, in his final years his only volitional movements confined to twitching muscles in his cheek. A computer interface allowed Hawking to use those movements to communicate and interact with the world. In the popular conscious-ness, he became an archetypally brilliant brain trapped in a non-functioning body. Mialet argued that he was in effect a brain hooked up to machinery: like Darth Vader, she said, he had become "more machine than man", that effect enhanced by Hawking's choice to retain his trademark retro "automated" voice.

Mialet's description drew intense criticism and condemnation, among others from the UK Motor Neurone Disease Association. But she didn't intend it as any kind of judgement, far less as insult. She wanted to encourage us to consider if we aren't all now part-machine, networked to our technologies and inhabiting worlds in which the real and the virtual bleed into one another. Because Hawking was at the same time so powerful in terms of the directive force of his thoughts – the twitch of a cheek was enough to activate his army of attendants and collaborators – and so powerless in terms of his physical agency, he embodied (so to speak) an extreme case, barely any different from the Machine Man of Ardathia or Olaf Stapledon's Fourth Men:

> his entire body and even his entire identity have become the property of a collective human-machine network. He is what I call a distributed centered-subject: a brain in a vat, living through the world outside the vat.

Comparing Hawking's situation with that of other individuals whose wishes are expressed and exerted through social and technological networks, Mialet concluded that "Someone who is powerful is a collective, and the more collective s/he becomes, the more singular they seem." The machinery of command becomes an extension of body and mind. Or to put it another way, the personhood of a "brain in a vat" might depend on its degree of agency.

Perhaps. But there is another, not incompatible, way to read Hawking's situation. The society in which he lived never, I think, quite came to terms with this man. We insisted on attributing to his fine scientific mind an almost preternatural genius, we heaped praise on his pleasantly wry but somewhat prosaic wit, we elected to sanctify – as much as secular society will sanction – a person of immense fortitude but of somewhat traditional, even conservative mien. In the end, excepting the small group of people close to him, Stephen Hawking's machine existence made us lose sight of the human – and to see only the legendary brain.

CHAPTER 8

RETURN OF THE MEATWARE

COMING TO TERMS
WITH OUR FLESHY SELVES

When science writer Carl Zimmer had his genome sequenced – every molecular "letter" of his DNA read and recorded – one of the scientists who helped him to understand what it meant indulged in the rhetorical flourish beloved of genomics researchers. Pointing at him, the scientist declared, "*That* is not Carl Zimmer." Then he pointed to the hard drive containing his genome data. "*This*," he announced, "is Carl Zimmer."

It wasn't (I hope) a comment designed for deep philosophical reflection, but its unguarded nature makes deconstruction all the more rewarding. We have been prepared for sentiments like this ever since the Human Genome Project – the international initiative that ran from 1990 to the early 2000s to read an entire human genome – was advertised as an effort to understand "what makes us human". The project, we were informed, was reading the book of life. It would reveal the blueprint of humankind. And so on.

What Zimmer was told puts that notion into its starkest expression: the body, the brain, the whole physical being of a person may be divested of personal identity and be reduced to a mere housing for, a husk of, the true individual. Our personhood itself, in this narrative, resides in abstract information that can be stored on the bytes of a computer disk. It is rather as if René Magritte were to reverse the message of his famous painting, insisting that this physical object in his hand were not a pipe at all because its essence of pipeness had been captured in the picture.

It is of course the genetic analogue of "downloading our brains": another strange and frantic flight from our fleshy nature. In this narrative, the body is de-authenticated and personhood becomes a computer code. You can put it in a book.

It's worth understanding how we got to this point. One reason, I believe, is the allure of the comprehensible. Human beings are frightfully complicated. To understand how the fertilized cell becomes an embryo, a fetus, a person, is a quest on which we have only just begun. Merely understanding how the body (usually) gets the right shape, let alone why some deviations from that shape are fatal while others remain viable, is a tremendous challenge. When we come to consider how the brain, the most complex object we know, becomes wired and orchestrates our behaviour from a mixture of contingent growth, sensory and environmental inputs, and genetic predisposition . . . well, then we struggle even to erect a multi-dimensional frame of reference for articulating the problem. The idea that all of this information can be reduced to an "instruction manual" encoded in four symbols that fits onto a hard drive promised to make the problem manageable. But it is an illusion.

There are other forces at play. If we believe that the real self can be put into a computer memory, we need no longer be troubled by corruption of the body. And of course mortification of and disgust for the flesh has a long and deep history in Western culture, where "carnal" carries a whiff of the debauched and morally corrupt.

Our meat-self is animal, while the cerebral – I mean metaphorically, not the icky grey matter inside our skull – elevates the human to a higher plane. Bodily disgust is quickly (and enthusiastically) taught and quickly learnt. Sex is a dirty necessity, even with our best efforts to sanctify bodily union via a socialized morality tightly policed by taboos. The same with death: grief is tinged with disgust and anxiety about decay, which our death rituals do their best to allay with embalmings and professionalization of bodily disposal. One of the worst (and best) horrors is the return of the dead, smelling of the charnel house.

Yes, when you think about the history of the body, it is no surprise at all that we have become so determined to find in modern biology permission to dematerialize the self into information. Such a self is timeless and perfectible – for information can be copied, preserved, even updated and edited (not to mention copyrighted and sold). Flesh is a mortal coil, but data is an eternal stream.

I don't think that anyone, faced with the warm, breathing person and the cold metallic hard drive, would have a moment's hesitation in deciding which is the real Carl Zimmer. He, like each of us, is not just made of flesh but is written into it. He is the fellow who enjoys a latte in New Haven's Book Trader café, not the binary code that allegedly encrypts those predilections and memories.

The temptation is to say we "inhabit" our bodies, even that we are "trapped in them". But that would be to collude with the fiction that the self is a thing that is housed, and perhaps that needs to be set free – that it could enjoy an independent existence, if only we can find the key to the exit. But we *are* our flesh, in the same sense that England is its hills and dales, its rivers and its cities and people. All of these things change over time; we never step onto the same England twice. The flux is part of the definition.

But even while we should reclaim our identity from the

genome-fantasists, we have to recognize that modern science complicates this insistence of returning to the body. For the body is far more plastic than we thought. I know that mine is, because I've seen some of it growing in an incubator across town, behaving as though it "wants" to become another superorganism like me, with thoughts of its own, searching with questing stumps of neural tissue for a body that is not there. My piece of flesh didn't get, could not yet be given, that opportunity. But it's by no means obvious that this is not, in principle, an option.

This new understanding of the capacities of human flesh disturbs old philosophical questions. Pretty much all of the traditional philosophies of self, from Descartes to Hume to Sydney Shoemaker, are predicated on the uniqueness and integrity of the individual.* But I seem now forced to accept the possibility (even if the actuality remains distant) that a piece of my arm could become an embryo, even without an act of conception: that pretty much every cell in my, or your, body is a potential person, or at least the progenitor gamete of a person. What does that mean for the religious ethics, not to mention the legalities, of the self? (If you can buy Elvis's sweat on eBay, does that mean you could "grow your own Elvis"?)

In the light of this protean ability of our flesh to grow, and keep growing, perhaps in any way we choose to guide it, how can we develop a moral framework for adjudicating life, death and identity? Where, in this seething mass of life, are we? Are we perhaps just a particular realization of some more general "essence of us" imbued in our living flesh?

What we seem to be seeking here is the root of our individuality.

---

* The most notable exception was the work of the late Derek Parfit, whose *Reasons and Persons* is becoming the starting point for philosophical reasoning about many of the implications of new technologies of people-making, from computer simulation to artificial reproductive technologies and cloning.

It may come as something of a shock to discover that biology fails to locate it.

*   *   *

John Donne's celebrated claim that "No man is an island" might be read retrospectively as a rearguard action, a spirited defence of the social self against early modernity's assertion of the independence and isolation of the individual. As Daniel Defoe's *Robinson Crusoe* – the first great fable of that modern view – illustrated, to be an island to oneself was to experience the contradictory terrors that you might never again enjoy human intercourse and that your self-sufficient and even comfortable isolation would be violated.

Pace Donne, our existence is totally and unavoidably solipsistic. We are trapped in our minds, ignorant of all that evades our senses and able only to infer, and never to experience, the minds of others. It is this (generally benign but potentially hellish) internal imprisonment that literature, art and some aspects of spirituality seek to alleviate. The mystery, though, is how the biology of mind and body conspire to assemble our illusion of unity. How are all our sensory inputs, each taking finite but differing times to process, integrated into the sensation that we are each a single conscious entity alive to the moment? The answer to that question still lies somewhere within the black box we label consciousness. We can only assume that the mind has evolved this capacity because it is a useful one: that perhaps a feeling of identity and individuality rewards us with a sense of agency, enabling us to make our way in the world.

What's odd, though, is that this should be necessary. We do not and presumably cannot know what it means to be a bat, as Thomas Nagel famously observed. But it's far harder to imagine what it is like to be a fungus, and it isn't clear that there is much meaning in asking what it is like to be a slime mould like *Dictyostelium discoideum*, which is both single-celled and multi-celled. It's doubtful there is anything much to be "like" when it comes to single cells. And yet these organisms make their own way in the world too, and with considerable success. There seems no reason to suppose that fungi, plants and bacteria, at least, have any need for "consciousness" in the way we know it: a feeling of individuality that goes beyond the ability to respond to stimuli in a self-preserving fashion.

Regardless of how or even why the human mind conjures the perception of individuality, it evidently does so. And this predisposes us to see it as a fact of nature. It stands to reason that there are also individual dogs and bats, and even individual cells. This singular existence is what the very word "organism" is supposed to invoke.

But even though science has sometimes been called "organized common sense", it can be frustratingly hard to make common sense scientific. When you really get down to it, the concept of the individual is all but impossible to define in scientific terms.

Our first instinct may be to define it with reference to the boundaries of the body. (Even bacteria have edges, right?) But my mini-brain put paid to that, as far as I'm concerned. If those tissues were not somehow a living piece of me, I do not know what they were. But it is clear in any case that my body is not just an assembly of human cells but a complex ecosystem composed of cells from hundreds of species. And as we've seen, the collective action of this community (especially but not exclusively the microbes in my gut) affects not only my metabolism but also my mood and other mental states – qualities that feel very much a part of the subjective "me". At any rate, "I" would be in a bad way without them.

This is the norm in nature. Symbiotic unions of cells from different species, often sharing metabolic and developmental duties, are everywhere. Plants acquire their nitrogen – a vital element for metabolism and growth – from the bacteria (rhizobia) in their root systems, where there are also fungi (mycorrhizae) to be found that carry out other essential functions. Corals depend on the nutrients provided by their symbiotic algae; some sponges contain 40 per cent by volume of bacteria.

"It seems," writes microbiologist James Shapiro, "that we need to think of *organism* as a term that has a much broader community-based or systemic meaning than the significance given by traditional perspectives based on the idea that each organism has its own separate, vertically inherited genome." It's no surprise that it would take a microbiologist to see beyond the traditional perspectives – for these folk are fixated precisely on life at the level of the cell and its environment, where the meanings of "life" become complicated. We could of course tell a nice inspirational story about all this "cooperation", to counteract the ruthless narrative of a "selfish" Darwinian struggle; but that seems neither essential nor particularly useful. Grant that random genetic variation subject to varying degrees of natural selection happens among proliferating cells, and evolution follows. Biology is what it is, and sometimes we should resist making it any more or less than that.

The ubiquity of symbiosis makes it clear that defining an individual organism in genetic terms doesn't generally make sense either. It will hopefully be clear now that the idea that our genome is the "essence of us" is fallacious from the outset. But it's not an essence of our organismic individuality either. Symbiosis means that we need things our genomes can't give us. As more and more species have their genomes sequenced, we are regularly encouraged to equate species with sequence, and sequence with organismal identity. But the genome of no complex organism contains all it needs for healthy growth and survival.

Even if our genome is not "complete", isn't it at least unique? That is the notion gene-sequencing companies push so hard, aiming to flatter: let us reveal the uniqueness of you. Of course, they know already that this will do nothing to flatter identical twins (quite the reverse, I'd have thought). Neither will it work for the many people who have mosaicism or chimerism in their cellular make-up (see page 78). And it certainly won't work for a definition of biological individuality in general, being irrelevant to bacteria that replicate asexually (and thus clonally) by cell division. No, there is no innate connection between the individuality of the organism and its genome – no one is going to propose, I think, that those cloned macaque monkeys are somehow "the same individual". Indeed, one of the weakest objections to human reproductive cloning (about which there are also some more persuasive concerns) is that it disrespects the "integrity" of a human being to give them the same genome as someone else. This is really just an expression of Scott Gilbert's myth of "DNA as soul".

Faced with these facts, some biologists have sought other ways to define biological individuality. How about the immune system, which after all is set up precisely to distinguish "self" (safe) tissues, cells and fragments thereof from "non-self" (potentially dangerous)? But immunity is not just protecting the organism from outside threats; it also looks for those coming from the inside, such as cancer cells, symbionts that get out of hand, even fetuses that the maternal body rejects. As Gilbert and his co-authors have written, the immune system is not merely the body's "armed forces" but also its "passport control": do you have the correct documentation to be here? "Immunity," they write, "does not merely guard the body against other hostile organisms in the environment; it also mediates the body's participation in a community of 'others' that contribute to its welfare."

In some desperation, a few researchers have tried to find a behavioural definition of the individual: it is that which is "autonomous" and "end-directed", acting towards some goal. But as usual in biology, introducing teleology causes more problems than it solves. Does the amoeba have a goal?

This might seem like pedantry. After all, we *are* individuals! But if that's so, philosopher Ellen Clarke has pointed out, then surely a biology worthy of the name ought to give some account of what it means. Individuals, she says, are "central to the inner logic of evolution by natural selection" – since it is the individual that evolution "sees", whose life or death determines whether the genes are passed on. But evidently, Clarke concludes, "we lack a theory telling us which lumps to count" in evolution and life.

This is because what "counts" (and what to count) depends on what questions we are asking and what scale we are observing. In short, there is not the slightest reason to suppose that what "counts" for our cells is what "counts" for us, any more than to think that what "matters" for quarks is what "matters" for planets. As a result, it should be no surprise that cell biology messes with our sense of self.

The best illustration I have seen of this fact is supplied by an organism none of us is in any hurry to encounter: the Portuguese man-of-war. This jellyfish has a nasty sting, its dangling tentacles loaded with a venom strong enough to kill a fish. But the Portuguese man-of-war is *not* actually a jellyfish at all. It looks pretty much indistinguishable from those organisms. Yet each "organism" is not an individual but a colony of small multi-celled animals called polyps. There are several "colonial animals" of this kind, collectively called zooids; corals are another example.

A Portuguese man-of-war: an "organism" that is truly a colony.

We are not like Portuguese men-of-war (if that is the plural); our cells are fully integrated, not apportioned into separate animals within the body. But that doesn't make us any the less a colony: a collaboration of cells, some of them human and some not, that has evolved this rather wonderful and mysterious tendency to consider itself a unique entity. Does the Portuguese man-of-war harbour, in its collective polyp-mind, any comparable illusion, I wonder?

* * *

Complicating the notion of individuality in biology is more than a nuisance for theories that seem to demand it. It pulls up at the roots the concept of nature that has grounded the discipline.

At one level, retaining the idea of the living world as a collection of individual, self-contained organisms is not only sensible but necessary – in much the same way as it is necessary for us to hold onto the idea of *objects* in a microscopic physical world imbued with quantum flux and devoid of hard edges. At the microscopic scale, we know that the large objects of classical physics – a book, a pen, an aircraft carrier – are fuzzy, their boundaries being a constant traffic of atoms. But it's worse than that, for quantum physics tells us that the subatomic particles making up matter can, in theory, pop up anywhere in the universe if we choose to look for them. It's just that some locations are much more likely than others. Efforts to make sense of quantum mechanics suggest that particles become entangled and interdependent once they interact, or that every particle "feels" all the others, or that particles and objects are constantly splitting into copies, or other things that make a nonsense of the clear-cut identities and edges of everyday perception. Yet if we don't maintain that fiction, not only does science become all but impossible but it fails in its job of offering a coherent description of our perceived reality.

So too with the biological individual: it would be foolish to abandon the concept. What we have to remember when we invoke it, though, is that we are doing so in order to formulate a picture of one level of the world. We are not voicing some fundamental truth. Indeed, if biology teaches us anything, it is that it is unwise to generalize about pretty much anything that is alive (including what we mean by that word). This is not because biology is lawless, but on the contrary because evolution is so central to it. Any way nature can "find" to impart a selective advantage is likely to be adopted, and it really does not matter what biologists have proclaimed to be allowed or forbidden: biology is no place for sacred cows or dogmas. If we truly believe in evolution, as of course we should, then we won't try to constrain it with supervening rules proclaiming this or that to be impossible. We will trust that it finds good strategies.

Darwinian evolution by natural selection amid random mutation is one of the main engines of biological evolution, and it makes sense of a great deal that we see in nature. But it doesn't make sense of everything. Evolution is driven by other things too: for example by random drift in which random mutations occur without selective pressure. Non-Darwinian change happens, as for example when some cells actively modify their genome (for example, its propensity to incur mutation) to improve survival and growth, or when cells that have been modified by some environmental influence pass on the modification to progeny cells. There is no reason to see these things as controversial or challenging, let alone as "disproving Darwin", for the simple reason that there is absolutely no "biological law" that would forbid them.

This is why narrative is so hazardous in biology. We seek simple stories – that genetic information flow is one-way, that genes are in a selfish struggle for survival, that genes make us who we are – because that is what we humans do. We can make them dogmas or laws if we like, but biology doesn't give a damn. It will do what suits it, and what it does is too complex, has too many exceptions, too great an inventive impulse, for any metaphor or narrative to fit perfectly.

So it is with biological individuality. If it proves to be a convenient fiction, that doesn't demolish biological ideas which draw on it (though it may well complicate them). Individuality is often a useful approximation at the level of large organisms like us – and no doubt that is why our brains have evolved to conceptualize the world in these terms. But it isn't a terribly useful notion for the cell. It is sometimes rightly said that if we were conscious electrons, we would have no concept of objects. Likewise, if we were conscious cells, we would not recognize individual organisms in the way that we humans conceive of them. We are not individuals all the way down.

Our cells are telling us this. It's worth listening to them. To use one of those hazardous human-shaped metaphors (but I think the cost is worth it): cells have a kind of wisdom, and we should allow ourselves to be humbled by it.

* * *

My little neural organoid was not really a "brain in a dish". There was never the slightest suggestion that it was conscious, or capable of anything that warranted being called cognition.

But just suppose we had been able to give it the vasculature and the developmental signals it would need to keep growing and to become even more truly brain-like. Or imagine we could grow neural cultures that resemble different parts of the brain and wire them together into a "brain assembloid". We still have no theory, or even a clear definition, of consciousness, but there is some reason to believe that it might arise in a specific region of the cortex. What if we could grow a brain organoid closely mimicking this region?

What kind of status should such a structure be afforded? Might it become capable of anything deserving to be called thought, or even reason? What would be the nature of its experience? And "who" would it be?

When I embarked on the Brains in a Dish project for Created Out of Mind, I considered that it would be foolish and presumptuous to ask such questions. As my mini-brain took shape, however, a paper was published in *Nature* by several leading neuroscientists and bioethicists who took the possibilities seriously. These things may be a long way off, the authors said, but we need to think about them now. They argued that "as brain surrogates become larger and more sophisticated, the possibility of them having capabilities akin to human sentience might become less remote." Such capacities "could include being able to feel (to some degree) pleasure, pain or distress; being able to store and retrieve memories; or perhaps even having some perception of agency or awareness of self." We need to think what a regulatory structure for such research should look like, says Hank Greely (one of that paper's authors) – and we have perhaps just 5 to 10 years before the questions become urgent.

Today's mini-brains have nothing like the complexity of the human brain. There are typically a mere 1 to 2 million neurons in their pea-sized mass, for example, compared to the 86 billion in the adult brain. What's more, the neurons in brain organoids are much less active, sending out signals ("firing") at just 3 to 4 per cent of the rate of those in the real brain and possessing little of their complexity of shape and structure.

But it's not absurd to entertain the idea of brain organoids with a degree of awareness. The numbers alone can be deceptive. For example, 80 per cent of the neurons in our brains are in the cerebellum, but it is possible for consciousness to develop even for a person lacking a cerebellum entirely – as was the case with at least one unfortunate Chinese woman born with that extreme developmental defect. And while brain neurons need sensory experience not only to build up a mental model of the world but to function properly at all, this could be supplied as an organoid grows. A brain organoid grown by one team of researchers, for example, began to develop a primitive kind of retina, and shining light on this area stimulated activity in the neurons. Madeline Lancaster has coupled the neurons of a brain organoid to muscle tissue and has seen muscle responses to the neural activity, in principle giving the organoid a capacity to affect and respond to its environment.

Christof Koch, another of the authors of the *Nature* commentary, questions the status of the mini-brain in rather confrontational terms: "We have to start thinking about it: is this thing in pain?" It's slightly shocking to realize that even experts don't know the answers to such questions. What can "pain" even mean in the absence of neuroreceptors through which pain is sensed by nerves in the body? (There are none of these receptors in the brain itself.) "Without knowing more about what consciousness is and what building blocks it requires, it might be hard to know what signals to look for in an experimental brain model," the *Nature* authors admit.

Who would determine – indeed, *what* would determine – the life or death of a sentient brain organoid? Will they need to have "guardians" appointed to look after their welfare, in the same way that individuals may be appointed for children involved in custody disputes? Would we owe sentient mini-brains a duty to provide a stimulating environment, meaningful memories and relationships? Do we have any idea what the notion of identity could mean to such a . . . thing? Entity? *Person*?

* * *

My own mini-brain has, however, had its day. Once Chris and Selina had grown it, they fixed it in formaldehyde, embedded it in a gel, and sectioned it for staining and imaging. I don't think I neglected any duty of pastoral care to this living entity, although I can't entirely shake off some lingering sentiment about the matter.

Is that the end of my adventure in the propagation of my own flesh beyond my all-too-mortal body? I wonder. "We still have your fibroblasts and your induced pluripotent stem cells in the liquid nitrogen," Chris tells me – "frozen in time, ready to be revived . . ."

ACKNOWLEDGEMENTS

"This book would not have been possible without . . ." might often seem like a piece of formulaic hyperbole, but it is certainly not in this case. I can safely say that I could not myself have grown a brain organoid from a piece of my arm, and until I first met Selina Wray of University College London I didn't even know it was possible. She and Chris Lovejoy have been immensely supportive, welcoming and generous in sharing their knowledge and skills about this astonishing process, and I am deeply grateful to them for starting me on the journey.

The impetus for the Brains in the Dish project came from artist Charlie Murphy, whose energy and imagination helped to see it through. Charlie's own responses to the experience, rendered mostly in glass objects of great delicacy and beauty, are inspiring. Thanks also to Ross Paterson for doing the dirty (in fact, of course, clinically clean) work of taking the biopsy.

There's also a "without whom . . ." element to the broader story. It was a huge pleasure and honour to be involved in Created Out of Mind, the residency at the Hub in the Wellcome Collection, London, for 2016–18. That ambitious project was led, with tremendous flair, patience, vision and good humour, by Seb Crutch of UCL, to whom I am full of gratitude for inviting me to be a part of the team working to change perceptions of dementia and to improve the lives and the care of people living with it. That team was a joy to work with, and included Caroline Evans, Kailey Nolan, Emilie Brotherhood, Janette Junghaus, Harriet Martin, Julian West, Paul Camic, Fergus Walsh,

Nick Fox, Gill Windle, Susanna Howard, Charlie Harrison, Hannah Zeilig, Millie van der Byl Williams, Tony Woods, Bridie Rollins and, oh goodness, I have probably forgotten others I should mention or who I never managed to meet, for which I ask their forgiveness.

I received advice from many experts in the course of preparing this book, as well as in writing some of the articles that informed it. Sometimes they will not have known (as I did not then know) that this is where their knowledge and wisdom would end up. Others kindly agreed to look at parts of the text and put right what I had wrong. Others supplied important materials for the research. They include Kristin Baldwin, Buzz Baum, Martin Birchall, Ali Brivanlou, Dan Davis, Sarah Franklin, Hank Greely, Ronald Green, Allon Klein, Christof Koch, Madeline Lancaster, Jennifer Lewis, Alison Murdoch, Werner Neuhausser, Brigitte Nerlich, Kathy Niakan, Andrew Reynolds, Adam Rutherford, Mitinori Saitou, Anil Seth, Marta Shahbazi, Deepak Srivastava, Azim Surani and Joseph Vacanti. Once again they have reminded me of how generous busy scientists and writers almost invariably are.

My editors in the UK and USA have given consistently smart advice, for which I owe many thanks to Hazel Eriksson, Myles Archibald and Karen Merikangas Darling. Karen obtained some thoughtful and helpful reviews whose anonymous authors I thank too. And I'm not sure what I would do without the support, encouragement and belief of my agent Clare Alexander, but it would probably involve finding another line of work.

Frankly, we're living through some grim times. The research and ideas I describe in this book create some alarming and disturbing possibilities, but they also speak – to me – of the determination of many folks to try to make things better, and of their ingenuity in that quest. It's never felt more vital, though, to receive support, companionship and love from my friends and family. I'm lucky to have that, and I am thankful.

Philip Ball
London, 2018

ENDNOTES

Prologue

"The slightly selfish big bit of chromosome": R. Dawkins, *The Selfish Gene*, 35. Oxford University Press, Oxford, 1976.

Chapter 1

"moved forward with a snake like motion": "Observationes D. Anthonii Lewenhoeck de Natis è semine genital Animalculis," *Philosophical Transactions of the Royal Society* **12**, 1040–1043, here 1041 (1677–78).

"There is one universal principle of development": T. Schwann, *Microscopic Researches into the Accordance in the Structure and Growth of Animals and Plants*, transl. H. Smith, 165. Sydenham Society, London, 1847.

"each cell is, within certain limits, an Individual": *ibid.* 2.

"elementary organism": E. W. von Brücke, "Die Elementarorganismen," *Sitzungsberichte der Kaierlichen Akademie Wien* **44**, 381–406 (1861).

"A cell . . . yes, that is really a person": Otis, 18 (2000).

"What the individual is on a grand scale": *ibid.* 21.

"for whatever the living cell is": E. B. Wilson, *The Cell in Development and Inheritance*, 13. Macmillan, New York, 1896.

"We must be careful": J. Gray, *A Textbook of Experimental Cytology*, 2. Cambridge University Press, Cambridge, 1931.

"The cell is making a particular kind of reappearance": Landecker, 4 (2007).

"the higher levels of order, form and function": Harold, 69 (2001).

"I prefer to think of the genome as akin": *ibid.* 69–70.

"Something is not accounted for very clearly": *ibid.* 65.

"DNA as our soul": Gilbert (2015).

Chapter 2

"the sperm fresh-sprung from the father's loins": Rosenfeld (1969).

"The parts of each organ help the other parts to form": Gilbert & Pinto-Correia, 91 (2017).

"an unfolding of pre-existing instructions": Keller, 21 (1995).

"greasy machines": *ibid.* 27.

"A more balanced and useful view": Nijhout, 444 (1990).

"We use words like *sister*": Zimmer, 384 (2018a).

First Interlude

"The higher animals, we learn": Reynolds (2008a).

"permanent ovum": *ibid.*

"our understanding of how powerful": Shapiro, 106 (2011).

Chapter 3

"In not just taking the animal body apart": Landecker, 67 (2007).

"rather limited value": Nicholas, 148 (1961).

"delicate surgical operation": Witkowski, 283 (1979).

"Carrel's new miracle points": Skloot, 68 (2010).

"In the next century, if infection, starvation": Friedman, 49 (2008).

"the creeping horror of the most morbid": Landecker, 92.

"If some day the scientists arrive": *ibid.* 98.

"bringing life under control": Wells, Huxley & Wells, 878 (1931).

"fragments of that eminent personage": *ibid.* 31.

"My mind went back to a day in 1918": Huxley (1926).

"I commend to the great public": *ibid.*

"the City during working hours": Wells, Huxley & Wells, 31.

"Doctor Farnham had a fleeting": Squier, 224 (2004).

"Now every separate part is tied": *ibid.* 79.

"Testicles and ovaries": *ibid.* 80.

"At the Strangeways lab a certain lad": *ibid.* 84.

"There is something rather romantic": *ibid.* 67.

"are responses to something otherwise not easily comprehended": Landecker, 142.

"living proof of the unexpected autonomy": *ibid.* 161.

"has secured for the patient": Skloot, 198.

"If allowed to grow uninhibited": *ibid.*

"a constant preoccupation with mass": Landecker, 174, 179.

"thrust into a kind of eternal life": *ibid.* 164.

"Tissues that move from bodies": Waldby & Mitchell, 34 (2006).

"If you repay me not on such a day": *The Merchant of Venice*, Act I, Scene iii.
"the human body exists beyond relations of commerce": Waldby & Mitchell, 23.
"deeply respected the rights of patients": https://agendapub.com/index.php/
   community/blog/105-the-curiouscase-of-john-moore-s-spleen
"What does it mean when the human body": Waldby & Mitchell, 7.
"And you know what?": O. Catts & I. Zurr, "Artists working with life (sciences)
   in contestable settings," *Interdisciplinary Science Reviews* **43**, 40–53, here
   47 (2018).
"the idea of *in vitro* production": *ibid.* 46.
"People . . . would be drawn to see the piece": *ibid.* 45.

Second Interlude
"It may be *triggered*": Davies, 137 (2019).
"Apparently the only thing our cells do": Raff, 121 (1998).
"we can't ignore the immune system": L. Lynch, public talk, "Schrödinger at
   75: the future of biology" Dublin, 6 September 2018.
"we are now using the taboo word": *ibid.*
"There is no such thing as a 'good microbe'": Yong, 80 (2016).

Chapter 4
"We thought at the time": Yamanaka (2012).
"to tell the truth, we did not expect that we had the answer": *ibid.*
"a paradigm shift in our understanding": https://www.nobelprize.org/nobel_
   prizes/medicine/laureates/2012/advanced-medicineprize2012.pdf
"Gene expression space can in some places": A. Klein, personal communication.
"It doesn't require any super-sophisticated bioengineering": Willyard, 521
   (2015).
"Simply by providing the right conditions": M. Lancaster, TEDxCERN talk,
   30 November 2015. https://www.youtube.com/watch?v=EjiWRINEatQ
"just what an embryo does": Madeline Lancaster, personal communication.
"You don't need a completely well formed human brain": Lancaster, personal
   communication.
"We hope to see the very earliest disease-associated changes": Selina Wray,
   personal correspondence.
"a technology of living substance": J. Loeb, letter to Ernst Mach, 26 February
   1890. In Pauly, 51 (1990).
"you're basically hacking the cell's code": D. Srivastava, personal communication.
"It appears that, in nature": Alvarado & Yamanaka, 115 (2014).

Chapter 5
"Isolated cells have the singular power": Carrel, 106 (1935).
"engendered by cells which": *ibid.* 107.
"an organ develops by means": *ibid.*
"somatic cells do not require the control": Wilson, 33 (2011).
"some means [of] artificial circulation": *ibid.* 32.
"We can take an ovary from a woman": Haldane, 64 (1924).
"if what has already been done": Anon. "Review of *Daedalus, or Science & the Future," Nature* **113**, 740 (1924).
"Already other parts of the human body": Burke, 3 (1938).
"it is now possible to construct": Rostand, 83 (1959).
"a crowd of various cells": Wilson, 31.
"we may find ourselves curing organoids": Marta Shahbazi, personal communication.
"I watched in agony and completely helpless": Vacanti, 397–398 (2007).
"It occurred to me that": *ibid.* 398.
"it became clear that, even with the very best": Martin Birchall, personal communication.
"I felt there must be a better": *ibid.*
"Considering how little we knew": *ibid.*
"the ideal material for building": Khademhosseini, Vacanti & Langer 68 (2009).
"If this research is allowed": Squier, 274–275 (2004).
"With every year I see more hope": Joseph Vacanti, personal communication.
"This is difficult science": *ibid.*
"the realization of Pygmalion's dream": Vladimir Mironov, personal communication.
"manufacturing platform for multicellular": Takebe *et al.* (2017).
"Animal-grown organs could transform": Conger (2018).
"to achieve his ambitious goal of flying": Rashid, Kobayashi & Nakauchi (2014).

Chapter 6
"If I'm honest": Franklin, Hopwood & Johnson, 17 (2009).
"Fertilized human eggs": Edwards, Bavister & Steptoe (1969).
"Embryos as we know them today": Morgan, 4 (2009).
"Had it not been for ectogenesis": Dronamraju, 84 (1995).
"There is no great invention": *ibid.* 36.
"What sort of creatures will these be?": Burke (1938).

"It will thus be seen": Wilson, 38 (2011).

"Next he had to get this life": Čapek (1921).

"There are the vats of liver": *ibid.*

"conquer the true human beings": Burke.

"a valuable technique with peculiar advantages": Wilson, 50.

"living tissues are growing": *ibid.*

"clearly not the sort of propaganda": *ibid.* 52.

"Tradition has been to regard": Rosenfeld, 47 (1969).

"the force of love may henceforth": *ibid.* 47, 49.

"reduce . . . the act of married love": P. Singer & D. Wells, *The Reproduction Revolution: New Ways of Making Babies*, 52. Oxford University Press, Oxford, 1984.

"She was born at around 11.47 pm": P. Gwynne, "All about that baby," *Newsweek* 7 August 1978, 44.

"Now when Rachel saw that she bore Jacob": Genesis, 30: 1–2.

"There is still not one social group": Gilbert & Pinto-Correia, 24–25 (2017).

"Remember that I am thy creature": Shelley, 68 (1818/2012).

"our social identities": Morgan, 62 (2009).

"Conception in vitro is now a normal": Franklin, 1 (2013a).

"confirms the viability": *ibid.* 308.

"is designed to precisely replicate": *ibid.* 234.

"is not that they are defying nature": van Dyck, 189–190 (1995).

"There is a revolutionary purpose": Franklin, 73 (2013a).

"good to think with": *ibid.* 29.

"Even the somewhat surprising scale": *ibid.* 148.

"Research on human embryos": Alison Murdoch, personal communication.

"complex amalgams of factual": M. Warnock *et al.*, "Report of the Committee of Inquiry into human fertilisation and embryology," HM Stationery Office, Para 11.9. London, 1984. Available at http://www.hfea.gov.uk/2068.html

"When I think about it": Azim Surani, personal communication.

"And the Lord God caused a deep": Genesis 2:21.

"regeneration of human gametes": Werner Neuhausser, personal communication.

"I get lots of emails": Surani, personal communication.

"Even with safeguards in place": Neuhausser, personal communication.

"If the human race as a whole": E. Dolgin, "Making babies: How to create human embryos with no egg or sperm," *New Scientist* 11 April 2018. https://www.newscientist.com/article/mg23831730-300-making-babies-how-to-create-human-embryos-with-no-egg-or-sperm/

"could blow away the biological barriers": A. Smajdor, talk at "Crossing frontiers: moving the boundaries of human reproduction," Progress Educational Trust, London, 8 December 2017. https://www.progress.org. uk/conference2017

"two people want their child": Greely, 190 (2016).

"evidence of just how wide-ranging": *ibid.*

Chapter 7

"a corpse would be re-animated": Shelley, 168 (1818/2012).

"I saw – with shut eyes": *ibid.*

"He sleeps, but he is awakened": *ibid.*

"supremely frightful would be the effect": *ibid.*

"In this the direct moral": *ibid.* 214.

"acts of damage-limitation": *ibid.* 415.

"rather study the established order": *ibid.* 236–237.

"They do say that man": Čapek (1921).

"Lonely island experiments": Friedman, 125 (2008).

"there is the distinct possibility": Rosenfeld, 44 (1969).

"there is no reason to believe": Pera (2017).

"It is unclear at which point a partial model": Rivron *et al.* (2018).

"if we had not had the prior experience": Alison Murdoch, personal communication.

"[He] foresees the day": D. Rorvik, *Brave New Baby*, 32. Doubleday, New York, 1971.

"could start us down a path towards": Lanphier *et al.* (2015).

"society will decide": Marchione (2018).

"The experiment was heedless": E. J. Topol, "Editing babies? We need to learn a lot more first," *New York Times* 27 November 2018. https://www. nytimes.com/2018/11/27/opinion/genetically-edited-babies-china.html

"It is imperative that the scientists": J. Doudna, *Berkeley News* 26 November 2018: https://news.berkeley.edu/2018/11/26/doudna-responds-to-claim-of-first-crispr-edited-babies/

"a global moratorium on all clinical uses": Lander *et al.*, 165 (2019).

"urgent need to confine the use": *ibid.*

"I believe that it will become": Ronald Green, personal communication.

"The creation of designer babies": Janssens (2018).

"They have a lot to be worried": Hank Greely, personal communication.

"Almost everything you can accomplish": *ibid.*

"The more eggs you can get": *ibid.*

"The science for safe and effective": *ibid.*

"Where there is a serious problem": Alto Charo, personal communication.

"our ability to love one another": *ibid.*

"we'll start seeing the use of gene editing": Ronald Green, personal communication.

"Genetic testing is a responsibility": Z. Corbyn, "'Genetic testing is a responsibility if you're having children,'" *The Observer* 8 January 2016. https://www.theguardian.com/science/2016/jan/08/anne-wojcicki-dna-genetics-testing-23andme-interview

"For better or worse": Green, personal communication.

"very different in size and temperament": Wilmut, Campbell & Tudge, 17 (2000).

"the very essence of humanity": J. Cohen (2018).

"around the world a modest number": Lauritzen (ed.), 114 (2001).

"cloning will have come to be looked on": *ibid.*

"the useless parts of the body": Bernal, 38 (1970).

"Instead of the present body structure": *ibid.* 39.

"Philosophies of life": More & Vita-More (2013).

"manipulation of the hereditary factors": Stapledon, 209 (1972).

"he was to be a normal human organism": *ibid.* 221.

"The sexual superwoman may be riddled": Ettinger (1972).

"Once that is done, sexual identity": More & Vita-More, 322.

"humanity itself is a disease": Ettinger, 4.

"Those who are willing to settle": *ibid.* preface (unnumbered).

"What you have made us is glorious": More & Vita-More, 449.

"the right to enhance one's body": *ibid.* 55.

Third Interlude

"If a heart could be kept beating": Squier, 219 (2004).

"This place is more terrible": *ibid.* 220.

"From now on, my pet": R. Dahl, "William and Mary," available at http://user.ceng.metu.edu.tr/~ucoluk/yazin/William_and_Mary.html

"When carried out under favorable conditions": https://alcor.org/FAQs/faq01.html

"the dominant myth of our age": Christof Koch, personal communication.

"Survival of your memories and personality": *ibid.*

"If you can bridge the gap": More & Vita-More, 164 (2013).

"old, weak, vulnerable, pitifully limited": *ibid.* 123.

"agents whose boundaries and components": *ibid.*

"utmost power and cunning": *The Philosophical Writings of Descartes*, Vol. II, transl. J. Cottingham, R. Stoothoff & D. Murdoch, 315. Cambridge University Press, Cambridge, 1984.

"I shall think that the sky": *ibid.*

"It might be suggested that you": G. Harman, *Thought*, 5. Princeton University Press, Princeton, 1973.

"If I am a Brain in a Vat, then I am not a Brain in a Vat": A. Brueckner, *Mind* **101**, 123–128 (1992).

"If I accept the argument": T. Nagel, *The View from Nowhere*, 73. Cambridge University Press, Cambridge, 1986.

"his entire body and even his entire identity": Mialet (2013).

"Someone who is powerful is a collective": *ibid.*

Chapter 8

"*That* is not Carl Zimmer": Zimmer (2018b).

"It seems that we need to think": Shapiro, 102 (2011).

"Immunity does not merely guard": Gilbert, Sapp & Tauber, 333 (2012).

"central to the inner logic": Clarke, 313 (2010).

"we lack a theory telling us": *ibid.* 323.

"as brain surrogates become larger": Farahany *et al.*, 430 (2018).

"Such capacities could include being able": *ibid.*

"We have to start thinking about it": A. Boyle, "Where does consciousness come from? Brain scientist closes in on the claustrum," *GeekWire* 3 November 2017. https://www.geekwire.com/2017/consciousness-come-brain-scientist-closes-claustrum/

"We still have your fibroblasts": Chris Lovejoy, personal communication.

BIBLIOGRAPHY

J. Aach, J. Lunshof, E. Iyer & G. M. Church, "Addressing the ethical issues raised by synthetic human entities with embryo-like features," *eLife* **6**, e20674 (2017).

A. S. Alvarado & S. Yamanaka, "Rethinking differentiation: stem cells, regeneration, and plasticity," *Cell* **157**, 110–119 (2014).

J. Andersen & S. P. Pasca, "Complementing the forebrain," *Nature* **563**, 44–45 (2018).

Anon., "Genome editing: proceed with caution," *The Lancet* **392**, 253 (2018).

C. Ariyachet *et al.*, "Reprogrammed stomach tissue as a renewable source of functional β cells for blood glucose," *Cell Stem Cell* **18**, 410–421 (2016).

T. Armstrong, *Modernism, Technology and the Body*. Cambridge University Press, Cambridge, 1998.

A. Atala, S. B. Bauer, S. Soker, J. J. Yoo & A. B. Retik, "Tissue-engineered autologous bladders for patients needing cystoplasty," *The Lancet* **367**, 1241–1246 (2006).

P. Ball, *Unnatural: The Heretical Idea of Making People*. Bodley Head, London, 2011.

P. Ball, "Self-repairing organs could save your life in a heartbeat," *New Scientist* 9 May 2018. https://www.newscientist.com/article/2168531-self-repairing-organs-could-save-your-life-in-a-heartbeat/

A. Banga, E. Akinci, L. V. Greder, J. R. Dutton & J. M. W. Slack, "In vivo reprogramming of Sox9⁺ cells in the liver to insulin-secreting ducts," *Proceedings of the National Academy of Sciences USA* **109**, 15336–15341 (2012).

S. Baruch, D. Kaufman & K. L. Hudson, "Genetic testing of embryos: practices and perspectives of US in vitro fertilization clinics," *Fertility and Sterility* **89**, 1053–1058 (2008).

L. Beccari, N. Moris, M. Girgin, D. A. Turner, P. Baillie-Johnson, A.-C. Cossy, M. P. Lutolf, D. Duboule & A. M. Arias, "Multi-axial self-organization properties of mouse embryonic stem cells into gastruloids," *Nature* **562**, 272–276 (2018).

Y. Belkaid & T. W. Hand, "Role of the microbiota in immunity and inflammation," *Cell* **157**, 121–141 (2014).

J. D. Bernal, *The World, the Flesh and the Devil: An Inquiry into the Future of the Three Enemies of the Rational Soul.* Jonathan Cape, London, 1970.

J. D. Biggers, "IVF and embryo transfer: historical origin and development," *Reproductive BioMedicine Online* **25**, 118–127 (2012).

M. J. Boland, J. L. Hazen, K. L. Nazor, A. R. Rodriguez, W. Gifford, G. Martin, S. Kupriyanov & K. K. Baldwin, "Adult mice generated from induced pluripotent stem cells," *Nature* **461**, 91–94 (2009).

S. Brenner, "Sequences and consequences," *Philosophical Transactions of the Royal Society B* **365**, 207–212 (2010).

J. Briscoe & S. Small, "Morphogen rules: design principles of gradient-mediated embryo patterning," *Development* **142**, 3996–4009 (2015).

N. Burke, "Could you love a chemical baby?," *Tit-Bits* 16 April 1938.

M. Caiazzo *et al.*, "Direct generation of functional dopaminergic neurons from mouse and human fibroblasts," *Nature* **476**, 224–227 (2011).

E. Callaway, "Second Chinese team reports gene editing in human embryos," https://www.nature.com/news/second-chinese-team-reports-gene-editing-in-human-embryos-1.19718 (8 April 2016).

E. Callaway, "Most popular human cell in science gets sequenced," *Nature News* 15 March 2013. https://www.nature.com/news/most-popular-human-cell-in-science-gets-sequenced-1.12609

J. Cao et al., "The single-cell transcriptional landscape of mammalian organogenesis", *Nature* **566**, 496-501 (2019).

K. Čapek, *R.U.R.*, transl. D. Wyllie. 1921. http://ebooks.adelaide.edu.au/c/capek/karel/rur

N. Carey, *The Epigenetics Revolution.* Icon, London, 2012.

A. L. Carlson, N. K. Bennett, N. L. Francis, A. Halikere, S. Clarke, J. C. Moore, R. P. Hart, K. Paradiso, M. Wernig, J. Kohn, Z. P. Pang & P. V. Moghe, "Generation and transplantation of human neurons in the brain using 3D microtopographic scaffolds," *Nature Communications* **7**, 10862 (2016).

A. Carrel, *Man, the Unknown.* Penguin, West Drayton, 1948.

A. N. Chang *et al.*, "Neural blastocyst complementation enables mouse forebrain organogenesis," *Nature* **563**, 126–129 (2018).

E. Clarke, "The problem of biological individuality," *Biological Theory* **5**, 312–325 (2010).

I. G. Cohen, "Disruptive reproductive technologies," *Science Translational Medicine* **9**, 10.1126/scitranslmed.aag2959 (2017).

J. Cohen, "An 'epic scientific misadventure': NIH head Francis Collins ponders fallout from CRISPR baby study," *Science* 30 November 2018. https://www.sciencemag.org/news/2018/11/epic-scientific-misadventure-nih-head-francis-collins-ponders-fallout-crispr-baby-study

M. A. Cohen, K. J. Wert, J. Goldmann, S. Markoulaki, Y. Buganim, D. Fu & R. Jaenisch, "Human neural crest cells contribute to coat pigmentation in interspecies chimeras after in utero injection into mouse embryos," *Proceedings of the National Academy of Sciences USA* **113**, 1570–1575 (2016).

K. Conger, "Growing human organs," *Stanford Medicine* **Winter** (2018). https://stanmed.stanford.edu/2018winter/caution-surrounds-research-into-growing-human-organs-in-animals.html

C. Crowley, M. Birchall & M. Seifalian, "Trachea transplantation: from laboratory to patient," *Journal of Tissue Engineering and Regenerative Medicine* **9**, 357–367 (2015).

D. Cyranoski, "'Reprogrammed' stem cells implanted into patient with Parkinson's," *Nature News* 14 November 2018, doi: 10.1038/d41586-018-07407-9

D. Cyranoski, "Egg engineers," *Nature* **500**, 392–394 (2013).

D. Cyranoski, "CRISPR-baby scientist fails to satisfy critics," *Nature* **564**, 13–14 (2018).

D. Cyranoski, "'Reprogrammed' stem cells to treat spinal-cord injuries for the first time", Nature News 22 February, doi: 10.1038/d41586-019-00656-2 (2019).

D. Cyranoski, "What's next for CRISPR babies?", *Nature* **566**, 440-442 (2019).

D. Cyranoski & H. Ledford, "Genome-edited baby claim provokes international outcry," *Nature* **563**, 607–608 (2018).

P. Davies, *The Demon in the Machine*. Allen Lane, London, 2019.

D. Davis, *The Beautiful Cure: Harnessing Your Body's Natural Defences*. Bodley Head, London, 2018.

R. L. Davis, H. Weintraub & A. B. Lassar, "Expression of a single transfected cDNA converts fibroblasts to myoblasts," *Cell* **51**, 987–1000 (1987).

A. De Los Angeles *et al.*, "Hallmarks of pluripotency," *Nature* **525**, 469–478 (2015).

A. Deglincerti, G. F. Croft, L. N. Pietilla, M. Zernicka-Goetz, E. D. Siggia & A. H. Brivanlou, "Self-organization of the *in vitro* attached human embryo," *Nature* **533**, 251–254 (2016).

S. Ding, "Deciphering therapeutic reprogramming," *Nature Medicine* **20**, 816–817 (2014).

W. F. Doolittle & S. L. Baldouf, "Origin and evolution of the slime molds (Mycetozoa)," *Proceedings of the National Academy of Sciences USA* **94**, 12007–12012 (1997).

K. R. Dronamraju (ed.), *Haldane's Daedalus Revisited*. Oxford University Press, Oxford, 1995.

A. D. Ebert, J. Yu, F. F. Rose Jr, V. B. Mattis, C. L. Lorson, J. A. Thomson & C. N. Svendsen, "Induced pluripotent stem cells from a spinal muscular atrophy patient," *Nature* **457**, 277–280 (2009).

R. G. Edwards, B. D. Bavister & P. C. Steptoe, "Early stages of fertilization *in vitro* of human oocytes matured *in vitro*," *Nature* **221**, 632–635 (1969).

M. Eiraku & Y. Sasai, "Self-formation of layered neural structures in three-dimensional culture of ES cells," *Current Opinion in Neurobiology* **22**, 768–777 (2012).

L. Eme, A. Spang, J. Lombard, C. W. Stairs & T. J. G. Ettema, "Archaea and the origin of eukaryotes," *Nature Reviews Microbiology* **15**, 711–723 (2017).

R. C. W. Ettinger, *Man into Superman*. St Martin's Press, New York, 1972.

N. A. Farahany *et al.*, "The ethics of experimenting with human brain tissue," *Nature* **556**, 429–432 (2018).

S. Franklin, *Biological Relatives: IVF, Stem Cells, and the Future of Kinship*. Duke University Press, Durham NC, 2013a.

S. Franklin, "Embryo watching: how IVF has remade biology," *Tecnoscienza* **4**, 23–43 (2013b).

S. Franklin, "Origin stories revisited: IVF as an anthropological project," *Culture, Medicine and Psychiatry* **30**, 547–555 (2006).

S. Franklin, "Rethinking reproductive politics in time, and time in UK reproductive politics: 1978–2008," *Journal of the Royal Anthropological Institute* **2014**, 109–125 (2014).

S. Franklin, "Revisiting reprotech: Firestone and the question of technology," in M. Merck & S. Sandford (eds), *Further Adventures of The Dialectic of Sex*, 29–60. Palgrave Macmillan, New York, 2010.

S. Franklin, "Conception through a looking glass: the paradox of IVF," *Reproductive BioMedicine Online* **27**, 747–755 (2013c).

S. Franklin, *Embodied Progress: A Cultural Account of Assisted Conception*. Routledge, London, 1997.

S. Franklin, N. Hopwood & M. Johnson (eds), *40 Years of IVF*, booklet to accompany a meeting at Christ's College Cambridge, 14 February 2009.

S. Franklin & H. Ragoné (eds), *Reproducing Reproduction: Kinship, Power, and Technological Innovation*. University of Pennsylvania Press, Philadelphia, 1998.

D. M. Friedman, *The Immortalists: Charles Lindbergh, Dr Alexis Carrel and Their Daring Quest to Live Forever*. JR Books, London, 2008.

L. Fu, X. Zhu, F. Yi, G. H. Liu & J. C. Izpisua Belmonte, "Regenerative medicine: transdifferentiation *in vivo*," *Cell Research* **24**, 141–142 (2014).

X. Gao, X. Wang & J. Chen, "*In vivo* reprogramming reactive glia into iPSCs to produce new neurons in the cortex following traumatic brain injury," *Scientific Reports* **6**, 22490 (2016).

S. F. Gilbert, "DNA as our soul: don't believe the advertising," *Huffington Post* 18 November 2015. https://www.huffingtonpost.com/scott-f-gilbert/dna-as-our-soul-believing_b_8590902.html

S. F. Gilbert, "A holobiont birth narrative: the epigenetic transmission of the human microbiome," *Frontiers in Genetics* **5**, article 282 (2014).

S. F. Gilbert, "Developmental biology, the stem cell of biological disciplines," *PLoS Biology* **15**, e2003691 (2017).

S. F. Gilbert, J. Sapp & A. I. Tauber, "A symbiotic view of life: we have never been individuals," *Quarterly Review of Biology* **87**, 325–341 (2012).

S. F. Gilbert (ed.), *A Conceptual History of Modern Embryology*. Plenum, New York, 1991.

S. Gilbert & C. Pinto-Correia, *Fear, Wonder, and Science in the New Age of Reproductive Biotechnology*. Columbia University Press, New York, 2017.

D. S. Glass & U. Alon, "Programming cells and tissues," *Science* **361**, 1199–1200 (2018).

H. T. Greely, *The End of Sex and the Future of Human Reproduction*. Harvard University Press, Cambridge Mass., 2016.

Z. Guo, L. Zhang, Z. Wu, Y. Chen, F. Wang & G. Chen, "In vivo direct reprogramming of reactive glial cells into functional neurons after brain injury and in an Alzheimer's disease model," *Cell Stem Cell* **14**, 188–202 (2014).

J. B. Gurdon, "The egg and the nucleus: a battle for supremacy," Nobel lecture 2012. https://www.nobelprize.org/prizes/medicine/2012/gurdon/lecture/

J. A. Hackett & M. A. Surani, "Regulatory principles of pluripotency: from the ground state up," *Cell Stem Cell* **15**, 416–430 (2014).

J. B. S. Haldane, *Daedalus, or Science & the Future*. Kegan Paul, Trench, Trubner & Co., London, 1924.

J. B. S. Haldane, *What Is Life?* Lindsay Drummond, London, 1949.

X. Han *et al.*, "Forebrain engraftment by human glial progenitor cells enhance synaptic plasticity and learning in adult mice," *Cell Stem Cell* **12**, 342–353 (2013).

D. Hanahan & R. A. Weinberg, "Hallmarks of cancer: the next generation," *Cell* **144**, 646–674 (2011).

F. M. Harold, *The Way of the Cell*. Oxford University Press, Oxford, 2001.

S. E. Harrison, B. Sozen, N. Christodoulou, C. Kyprianou & M. Zernicka-Goetz, "Assembly of embryonic and extra-embryonic stem cells to mimic embryogenesis in vitro," *Science* eaal1810 (2017).

K. Hayashi, S. Ogushi, K. Kurimoto, S. Shimamoto, H. Ohta & M. Saitou, "Offspring from oocytes derived from in vitro primordial germ cell-like cells in mice," *Science* **338**, 971–975 (2012).

K. Hayashi, H. Ohta, K. Kurimoto, S. Aramaki & M. Saitou, "Reconstitution of the mouse germ cell specification pathway in culture by pluripotent stem cells," *Cell* **146**, 519–532 (2011).

K. K. Hirschi, S. Li & K. Roy, "Induced pluripotent stem cells for regenerative medicine," *Annual Reviews of Biomedical Engineering* **16**, 277–294 (2014).

N. Hopwood, "'Giving body' to embryos: modelling, mechanism, and the microtome in late nineteenth-century anatomy," *Isis* **90**, 462–496 (1999).

N. Hopwood, "Producing development: the anatomy of human embryos and the norms of Wilhelm His," *Bulletin of the History of Medicine* **74**, 29–79 (2000).

J. Huxley, "The tissue-culture king," *Cornhill Magazine* **60**, 422–458 (1926). Available at http://www.revolutionsf.com/fiction/tissue/

I. Hyun, A. Wilkerson & J. Johnston, "Embryology policy: revisit the 14-day rule," *Nature* **533**, 169–171 (12 May 2016).

A. C. J. W. Janssens, "Those designer babies everyone is freaking out about – it's not likely to happen," *The Conversation* 10 December 2018. https://theconversation.com/those-designer-babies-everyone-is-freaking-out-about-its-not-likely-to-happen-103079

C. Y. Johnson, "Lab-grown brain bits open windows to the mind – and a maze of ethical dilemmas," *Washington Post* 2 September 2018.

N. L. Jorstad, M. S. Wilken, W. N. Grimes, S. G. Wohl, L. S. VandenBosch, T. Yoshimatsu, R. O. Wong, F. Rieke & T. A. Reh, "Stimulation of functional neuronal regeneration from Müller glia in adult mice," *Nature* **548**, 103–107 (2017).

E. F. Keller, *Refiguring Life: Metaphors of Twentieth-Century Biology*. Columbia University Press, New York, 1995.

A. Khademhosseini, J. P. Vacanti & R. Langer, "Progress in tissue engineering," *Scientific American* **300**, 64–71 (2009).

T. Kikuchi *et al.*, "Human iPS cell-derived dopaminergic neurons function in a primate Parkinson's disease model," *Nature* **548**, 592–596 (2017).

G. J. Knott & J. A. Doudna, "CRISPR-Cas guides the future of genetic engineering," *Science* **361**, 866–869 (2018).

P. Koch & J. Ladewig, "A little bit of guidance: mini brains on their route to adolescence," *Cell Stem Cell* **21**, 157–158 (2017).

D. B. Kolesky, K. A. Human, M. A. Skylar-Scott & J. A. Lewis, "Three-dimensional bioprinting of thick vascularized tissues," *Proceedings of the National Academy of Sciences USA* **113**, 3179–3184 (2016).

D. B. Kolesky, R. L. Truby, A. S. Gladman, T. A. Busbee, K. A. Homan & J. A. Lewis, "3D bioprinting of vascularized, heterogeneous cell-laden tissue constructs," *Advanced Materials* **26**, 3124–3130 (2014).

J. Ladewig, P. Koch & O. Brüstle, "Leveling Waddington: the emergence of direct programming and the loss of cell fate hierarchies," *Nature Reviews Molecular Cell Biology* **14**, 225–236 (2013).

J. Lambert, "Should evolution treat our microbes as part of us?," *Quanta* 20 November 2018. https://www.quantamagazine.org/should-evolution-treat-our-microbes-as-part-of-us-20181120/

M. A. Lancaster & J. A. Knoblich, "Organogenesis in a dish: modeling development and disease using organoid technologies," *Science* **345**, 283 and supplement 1247125 (2014).

M. A Lancaster, M. Renner, C.-A. Martin, D. Wenzel, L. S. Bicknell, M. E. Hurles, T. Homfray, J. M. Penninger, A. P. Jackson & J. A. Knoblich, "Cerebral organoids model human brain development and microcephaly," *Nature* **501**, 373–379 (2013).

H. Landecker, *Culturing Life: How Cells Became Technologies.* Harvard University Press, Cambridge Mass., 2007.

E. Lander *et al.*, "Adopt a moratorium on heritable genome editing," *Nature* **567**, 165–168 (2019).

E. Landhuis, "Tapping into the brain's star power," *Nature* **563**, 141–143 (2018).

N. Lane, *Life Ascending: The Ten Great Inventions of Evolution.* Profile, London, 2009.

E. Lanphier, F. Urnov, S. E. Haecker, M. Werner & J. Smolenski, "Don't edit the human germ line," *Nature* **519**, 410–411 (2015).

P. Lauritzen (ed.), *Cloning and the Future of Human Embryo Research.* Oxford University Press, New York, 2001.

H. Ledford, "CRISPR fixes disease gene in viable human embryos," *Nature* **548**, 13–14 (2017).

P. Li, H. Hu, S. Yang, R. Tian, Z. Zhang, W. Zhang, M. Ma, Y. Zhu, X. Guo, Y. Huang, Z. He & Z. Li, "Differentiation of induced pluripotent stem cells into male germ cells *in vitro* through embryoid body formation and retinoic acid or testosterone induction," *BioMed Research International* doi:10.1155/2013/608728 (2013).

Y.-C. Li, K. Zhu & T.-H. Young, "Induced pluripotent stem cells, from *in vitro* tissue engineering to *in vivo* allogenic transplantation," *Journal of Thoracic Disease* **9**, 455–459 (2017).

S. Lidgard & L. K. Nyhart (eds), *Biological Individuality: Integrating Scientific, Philosophical, and Historical Perspectives.* University of Chicago Press, Chicago, 2017.

M. Lie, "Reproduction inside/outside: medical imaging and the domestication of assisted reproductive technologies," *European Journal of Women's Studies* **22**, 53–69 (2015).

M.-L. Liu, T. Zang & C.-L. Zhang, "Direct lineage reprogramming reveals disease-specific phenotypes of motor neurons from human ALS patients," *Cell Reports* **14**, 1–14 (2016).

M.-L. Liu, T. Zang, Y. Zou, J. C. Chang, J. R. Gibson, K. M. Huber & C.-L. Zhang, "Small molecules enable neurogenin 2 to efficiently convert human fibroblasts into cholinergic neurons," *Nature Communications* **4**, 2183 (2013).

T.-Y. Lu, B. Lin, J. Kim, M. Sullivan, K. Tobita, G. Salama & L. Yang, "Repopulation of decellularized mouse heart with human induced pluripotent stem cell-derived cardiovascular progenitor cells," *Nature Communications* **4**, 2307 (2013).

S. Luo *et al.*, "Divergent lncRNAs regulate gene expression and lineage differentiation in pluripotent cells," *Cell Stem Cell* **18**, 637–652 (2016).

H. Ma *et al.*, "Correction of a pathogenic gene mutation in human embryos," *Nature* **548**, 413–419 (2017).

J. Maierschein, *Whose View of Life? Embryos, Cloning, and Stem Cells.* Harvard University Press, Cambridge Mass., 2003.

M. Marchione, "Chinese researcher claims first gene-edited babies," *AP News* 26 November 2018. https://www.apnews.com/4997bb7aa36c45449b488e19ac83e86d

E. Martin, "The egg and the sperm: how science has constructed a romance based on stereotypical male–female roles," *Journal of Women in Culture and Society* **16**, 485–501 (1991).

W. Martin & E. V. Koonin, "Introns and the origin of nucleus-cytosol compartmentalization," *Nature* **440**, 41–45 (2006).

I. Martyn, T. Y. Kanno, A. Ruzo, E. D. Siggia & A. H. Brivanlou, "Self-organization of a human organizer by combined Wnt and Nodal signaling," *Nature* **558**, 132–135 (2018).

H. Masumoto & J. K. Yamashita, "Human iPS cell-derived cardiac tissue sheets: a platform for cardiac regeneration," *Current Treatment Options in Cardiovascular Medicine* **18**, 65 (2016).

K. S. Matlin, J. Maienschein & M. D. Laubichler, *Visions of Cell Biology.* University of Chicago Press, Chicago, 2018.

H. Matsunari *et al.*, "Blastocyst complementation generates exogenic pancreas in vivo in apancreatic cloned pigs," *Proceedings of the National Academy of Sciences USA* **110**, 4557–4562 (2013).

T. Matsuo, H. Masumoto, S. Tajima, T. Ikuno, S. Katayama, K. Minakata, T. Ikeda, K. Yamamizu, Y. Tabata, R. Sakata & J. K. Yamashita, "Efficient long-term survival of cell grafts after myocardial infarction with thick viable cardiac tissue entirely from pluripotent stem cells," *Scientific Reports* **5**, 16842 (2015).

P. Mazzarello, "A unifying concept: the history of cell theory," *Nature Cell Biology* **1**, E13–E15 (1999).

K. W. McCracken, E. M. Catá, C. M. Crawford, K. L. Sinagoga, M. Schumacher, B. E. Rockich, Y.-H. Tsai, C. N. Mayhew, J. R. Spence, Y. Zavros & J. M. Wells, "Modelling human development and disease in pluripotent stem-cell-derived gastric organoids," *Nature* **516**, 400–404 (2014).

P. B. Medawar, *The Uniqueness of the Individual*. Methuen, London, 1957.

H. Mialet, "On Stephen Hawking, Vader, and being more machine than man," *Wired* 8 January 2013. https://www.wired.com/2013/01/hawking-machine-man-robots/

C. C. Miranda, T. G. Fernandes, M. M. Diogo & J. M. S. Cabral, "Towards multi-organoid systems for drug screening and applications," *Bioengineering* **5**, E49 (2018).

M. More & N. Vita-More (eds), *The Transhumanist Reader*. Wiley-Blackwell, Chichester, 2013.

L. Morgan, "Embryo tales," in S. Franklin & M. Lock (eds), *Remaking Life and Death: Toward an Anthropology of the Biosciences*, 261–291. School of American Research Press, Santa Fe NM, 2003.

L. M. Morgan, *Icons of Life: A Cultural History of Human Embryos*. University of California Press, Berkeley, 2009.

S. A. Morris, "Human embryos cultured *in vitro* to 14 days," *Royal Society Open Biology* **7**, 170003 (2017).

C. Mummery, I. Wilmut, A. van de Stolpe & B. A. J. Roelen, *Stem Cells: Scientific Facts and Fiction*. Academic Press, London, 2011.

S. V. Murphy & A. Atala, "3D bioprinting of tissues and organs," *Nature Biotechnology* **8**, 773–785 (2014).

J. S. Nicholas, "Ross Granville Harrison 1870–1959," *Biographical Memoirs of the National Academy of Sciences*. National Academy of Sciences, Washington, DC, 1961.

H. F. Nijhout, "Metaphors and the role of genes in development," *Bioessays* **12**, 441–446 (1990).

Nuffield Council on Bioethics, *Genome Editing and Human Reproduction: Social and Ethical Issues*. Nuffield Council on Bioethics, London, 2018.

P. Nurse, "Life, logic and information," *Nature* **454**, 424–426 (2008).

H. Okae, H. Toh, T. Sato, H. Hiura, S. Takahashi, K. Shirane, Y. Kabayama, M. Suyama, H. Sasaki & T. Arima, "Derivation of human trophoblast stem cells," *Cell Stem Cell* **22**, 50–63 (2018).

L. Otis, *Membranes: Metaphors of Invasion in Nineteenth-Century Literature, Science, and Politics*. Johns Hopkins University Press, Baltimore, 2000.

F. W. Pagliuca, J. R. Millman, M. Gürtler, M. Segel, A. Van Dervort, J. H.

Ryu, Q. P. Peterson, D. Greiner & D. A. Melton, "Generation of functional human pancreatic β cells in vitro," *Cell* **159**, 428–439 (2014).

S. P. Pasca, "Assembling human brain organoids," *Science* **363**, 126–127 (2019).

P. J. Pauly, *Controlling Life: Jacques Loeb and the Engineering Ideal in Biology.* University of California Press, Berkeley, 1990.

M. Pera, "Embryogenesis in a dish," *Science* **356**, 137–138 (2017).

M. F. Pera, "Human embryo research and the 14-day rule," *Development* **144**, 1923–1925 (2017).

U. Pfisterer, A. Kikeby, O. Torper, J. Wood, J. Nelander, A. Dufour, A. Björklund, O. Lindvall, J. Jakobsson & M. Parmar, "Direct conversion of human fibroblasts to dopaminergic neurons," *Proceedings of the National Academy of Sciences USA* **108**, 10343–10348 (2011).

B. Pijuan-Sala et al., "A single-cell molecular map of mouse gastrulation and early organogenesis", *Nature* **566**, 490-495 (2019).

R. Plomin, *Blueprint: How DNA Makes Us Who We Are.* Allen Lane, London, 2018.

J. Pollak, M. S. Wilken, Y. Ueki, K. E. Cox, J. M. Sullivan, R. J. Taylor, E. M. Levine & T. A. Reh, "ASC1 reprograms mouse Müller glia into neurogenic retinal progenitors," *Development* **140**, 2619–2631 (2013).

J. Qiu, "Chinese government funding may have been used for 'CRISPR babies' project, document suggests", *STAT News* 25 February 2019. https://www. statnews.com/2019/02/25/crispr-babies-study-china-government-funding/

M. C. Raff, "Social controls on cell survival and cell death," *Nature* **356**, 397–400 (1992).

M. C. Raff, "Cell suicide for beginners," *Nature* **396**, 119–122 (1998).

T. Rashid, T. Kobayashi & H. Nakauchi, "Revisiting the flight of Icarus: making human organs from PSCs with large animal chimeras," *Cell Stem Cell* **15**, 406–409 (2014).

A. Regalado, "A new way to reproduce," *MIT Technology Review* 7 August 2017. https://www.technologyreview.com/s/608452/a-new-way-to-reproduce/

A. S. Reynolds, "The redoubtable cell," *Studies in History and Philosophy of the Biological and Biomedical Sciences* **41**, 194–201 (2010).

A. S. Reynolds, "Haeckel and the theory of the cell-state: remarks on the history of a bio-political metaphor," *History of Science*, **Summer, xlvi**, 123–152 (2008b).

A. S. Reynolds, "Amoebae as exemplary cells: the protean nature of an elementary organism," *Journal of the History of Biology* **41**, 307–337 (2008a).

A. S. Reynolds, "The cell's journey: from metaphorical to literal factory," *Endeavour: A quarterly magazine reviewing the history and philosophy of science in the service of mankind* **31**, 65–70 (2007).

A. S. Reynolds, "The theory of the cell state and the question of cell autonomy in nineteenth and early-twentieth century biology," *Science in Context* **20**, 71–95 (2007).

A. S. Reynolds, *The Third Lens: Metaphor and the Creation of Modern Cell Biology*. University of Chicago Press, Chicago, 2018.

A. S. Reynolds & N. Huelsmann, "Ernst Haeckel's discovery of Magosphaera planula: a vestige of metazoan origins?" *History and Philosophy of the Life Sciences* **30**, 339–386 (2008).

A. Rezania, J. E. Bruin, P. Arora, A. Rubin, I. Batushanksy, A. Asadi, S. O'Dwyer, N. Quiskamp, M. Mojibian, T. Albrecht, Y. H. Yang, J. D. Johnson & T. J. Kieffer, "Reversal of diabetes with insulin-producing cells derived in vitro from human pluripotent stem cells," *Nature Biotechnology* **32**, 1121–1133 (2014).

N. Rivron *et al.*, "Debate ethics of embryo models from stem cells," *Nature* **564**, 183–185 (2018).

R. E. Rodin & C. A. Walsh, "Somatic mutation in pediatric neurological diseases," *Pediatric Neurology* 10.1016/j.pediatrneurol.2018.08.008 (2018).

A. Rosenfeld, "Challenge to the miracle of life," *Life* 13 June 1969, 38–51.

J. Rostand, *Can Man Be Modified?*, transl. J. Griffin. Basic Books, New York, 1959.

M. Saito & H. Miyauchi, "Gametogenesis from pluripotent stem cells," *Cell Stem Cell* **18**, 721–735 (2016).

Y. Sasai, "Next-generation regenerative medicine: organogenesis from stem cells in 3D culture," *Cell Stem Cell* **12**, 520–530 (2013).

T. Sato, K. Katagiri, T. Yokonishi, Y. Kubota, K. Inoue, N. Ogonuki, S. Matoba, A. Ogura & T. Ogawa, "*In vitro* production of fertile sperm from murine spermatogonial stem cell lines," *Nature Communications* **2**, 472 (2011).

A. K. Seth & M. Tsakiris, "Being a beast machine: the somatic basis of self-hood," *Trends in Cognitive Science* **22**, 969–981 (2018).

M. N. Shahbazi & M. Zernicka-Goetz, "Deconstructing and reconstructing the mouse and human early embryo," *Nature Cell Biology* **20**, 878–887 (2018).

M. N. Shahbazi *et al.*, "Self-organization of the human embryo in the absence of maternal tissues," *Nature Cell Biology* **18**, 700–708 (2016).

Y. Shao, K. Taniguchi, R. F. Townshend, T. Miki, D. L. Gumucio & J. Fu, "A pluripotent stem cell-based model for post-implantation human amniotic sac development," *Nature Communications* **8**, 208 (2017).

J. A. Shapiro, *Evolution: A View from the 21st Century*. FT Press, Upper Saddle River NJ, 2011.

M. Shelley, *Frankenstein*. Second Norton Critical Edition, ed. J. P. Hunter. W. W. Norton, New York, 2012.

H. Shen, "Embryo assembly 101," *Nature* **559**, 19–22 (2018).

M. Simunovic & A. H. Brivanlou, "Embryoids, organoids and gastruloids: new approaches to understanding embryogenesis," *Development* **144**, 976–985 (2017).

V. K. Singh, M. Kalsan, N. Kumar, A. Saini & R. Chandra, "Induced pluripotent stem cells: applications in regenerative medicine, disease modeling, and drug discovery," *Frontiers in Cell and Developmental Biology* **3**, Article 2 (2015).

A. Skardal, T. Shupe & A. Atala, "Organoid-on-a-chip and body-on-a-chip systems for drug screening and disease modeling," *Drug Discovery Today* **21**, 1399–1411 (2016).

R. Skloot, *The Immortal Life of Henrietta Lacks*. Macmillan, London, 2010.

B. Sozen, G. Amadei, A. Cox, R. Wang, E. Na, S. Czukiewska, L. Chappell, T. Voet, G. Michel, N. Jing, D. M. Glover & M. Zernicka-Goetz, "Self-assembly of embryonic and two extra-embryonic stem cell types into gastrulating embryo-like structures," *Nature Cell Biology* **20**, 979–989 (2018).

S. M. Squier, *Babies in Bottles: Twentieth-Century Visions of Reproductive Technology*. Rutgers University Press, New Brunswick NJ, 1994.

S. M. Squier, *Liminal Lives: Imagining the Human at the Frontiers of Biomedicine*. Duke University Press, Durham NC, 2004.

D. Srivastava & N. DeWitt, "In vivo cellular reprogramming: the next generation," *Cell* **166**, 1386–1396 (2016).

M. Stadtfeld & K. Hochedlinger, "Induced pluripotency: history, mechanisms, and applications," *Genes and Development* **24**, 2239–2263 (2010).

O. Stapledon, *Last and First Men/Last Men in London*. Penguin, Harmondsworth, 1972.

P. C. Steptoe, R. G. Edwards & J. M. Purdy, "Human blastocysts grown in culture," *Nature* **229**, 132–133 (1971).

Z. Su, W. Niu, M.-L. Liu, Y. Zou & C.-L. Zhang, "*In vivo* conversion of astrocytes to neurons in the injured adult spinal cord," *Nature Communications* **5**, 3338 (2014).

K. Takahashi & S. Yamanaka, "Induced pluripotent stem cells in medicine and biology," *Development* **140**, 2457–2461 (2013).

N. Takata & M. Eiraku, "Stem cells and genome editing: approaches to tissue regeneration and regenerative medicine," *Journal of Human Genetics* **63**, 165–178 (2018).

T. Takebe et al., "Massive and reproducible production of liver buds entirely from human pluripotent stem cells," *Cell Reports* **21**, 2661–2670 (2017).

B. Tasic et al., "Shared and distinct transcriptomic cell types across neocortical areas," *Nature* **563**, 72–78 (2018).

M. Y. Turco et al., "Trophoblast organoids as a model for maternal-fetal interactions during human placentation," *Nature* https://www.nature.com/articles/s41586-018-0753-3 (2018).

J. P. Vacanti, "Tissue engineering and regenerative medicine," *Proceedings of the American Philosophical Society* **151**, 395–402 (2007).

S. C. van den Brink, P. Baillie-Johnson, T. Balayo, A.-K. Hadjantonakis, S. Nowotschin, D. A. Turner & A. M. Arias, "Symmetry breaking, germ layer specification and axial organization in aggregates of mouse embryonic stem cells," *Development* **141**, 4231–4242 (2014).

J. van Dyck, *Manufacturing Babies and Public Consent*. Macmillan, Basingstoke, 1995.

T. Vierbuchen, A. Ostermeier, Z. P. Pang, Y. Kokubu, T. C. Südhof & M. Wernig, "Direct conversion of fibroblasts to functional neurons by defined factors," *Nature* **463**, 1035–1041 (2010).

G. Vogel, "Human organs grown in pigs? Not so fast," *Science* 26 January 2017. http://www.sciencemag.org/news/2017/01/human-organs-grown-pigs-not-so-fast

D. E. Wagner, C. Weinreb, Z. M. Collins, J. A. Briggs, S. G. Megason & A. M. Klein, "Single-cell mapping of gene expression landscapes and lineage in the zebrafish embryo," *Science* 10.1126/science.aar4362 (2018).

C. Waldby & R. Mitchell, *Tissue Economies: Blood, Organs, and Cell Lines in Late Capitalism*. Duke University Press, Durham NC, 2006.

A. Warmflash, B. Sorre, F. Etoc, E. D. Siggia & A. H. Brivanlou, "A method to recapitulate early embryonic spatial patterning in human embryonic stem cells," *Nature Methods* **11**, 847–854 (2014).

R. Weinberg, *One Renegade Cell: The Quest for the Origins of Cancer*. Weidenfeld & Nicolson, London, 1998.

R. Weinberg, "Coming full circle – from endless complexity to simplicity and back again," *Cell* **157**, 267–271 (2014).

D. J. Weiss, M. Elliott, Q. Jang, B. Poole & M. Birchall, "Tracheal bioengineering: the next steps. Proceeds of an International Society of Cell Therapy Pulmonary Cellular Therapy Signature Series Workshop, Paris, France. April 22, 2014," *Cytotherapy* **16**, 1601–1613 (2014).

H. G. Wells, J. Huxley & G. P. Wells, *The Science of Life*. Cassell, London, 1931.

C. Willyard, "Rise of the organoids," *Nature* **523**, 520–522 (2015).

I. Wilmut, K. Campbell & C. Tudge, *The Second Creation*. Headline, London, 2000.

D. Wilson, *Tissue Culture in Science and Society*. Palgrave Macmillan, London, 2011.

J. A. Witkowski, "Alexis Carrel and the mysticism of tissue culture," *Medical History* **23**, 279–296 (1979).

C. R. Woese, "A new biology for a new century," *Microbiology and Molecular Biology Reviews* **68**, 173–186 (2004).

J. Wu *et al.*, "Interspecies chimerism with mammalian pluripotent stem cells," *Cell* **168**, 473–486 (2017).

J. Wu, H. T. Greely, R. Jaenisch, H. Nakauchi, J. Rossant & J. C. Izpisua Belmonte, "Stem cells and interspecies chimaeras," *Nature* **549**, 51–59 (2016).

Y.-Y. Wu, F.-L. Chiu, C.-S. Yeh & H.-C. Kuo, "Opportunities and challenges for the use of induced pluripotent stem cells in modeling neurodegenerative disease," *Open Biology* **8**, 180177 (2019).

J. Xu, Y. Du & H. Deng, "Direct lineage reprogramming: strategies, mechanisms, and applications," *Cell Stem Cell* **16**, 119–134 (2015).

P.-F. Xu, N. Houssin, K. F. Ferri-Lagneau, B. Thisse & C. Thisse, "Construction of a vertebrate embryo from two opposing morphogen gradients," *Science* **344**, 87–89 (2014).

S. Yamanaka, "The winding road to pluripotency," Nobel lecture 2012. https://www.nobelprize.org/prizes/medicine/2012/yamanaka/lecture/

C. Yamashiro, K. Sasaki, Y. Yabuta, Y. Kojima, T. Nakamura, I. Okamoto, S. Yokobayashi, Y. Murase, Y. Ishikura, K. Shirna, H. Sasaki, T. Yamamoto & M. Saitou, "Generation of human oogonia from induced pluripotent stem cells in vitro," *Science* 10.1136/science.aat1674 (2018).

D. Yates, "Reprogramming the residents," *Nature Reviews Neuroscience* **14**, 739 (2013).

E. Yong, *I Contain Multitudes*. Random House, London, 2016.

E. Yong, "A reckless and needless use of gene editing on human embryos," *The Atlantic* 26 November 2018. https://www.theatlantic.com/science/archive/2018/11/first-gene-edited-babies-have-allegedly-been-born-in-china/576661/

E. Yong, "The CRISPR baby scandal gets worse by the day," *The Atlantic* 3 December 2018. https://www.theatlantic.com/science/archive/2018/12/15-worrying-things-about-crispr-babies-scandal/577234/

R. Zhang, P. Han, H. Yang, K. Ouyang, D. Lee, Y.-F. Lin, K. Ocorr, G. Kang, J. Chen, D. Y. R. Stainier, D. Yelon & N. C. Chi, "*In vivo* cardiac reprogramming contributes to zebrafish heart regeneration," *Nature* **498**, 497–501 (2013).

Q. Zhou, J. Brown, A. Kanarek, J. Rajagopal & D. A. Melton, "*In vivo* reprogramming of adult pancreatic exocrine cells to β-cells," *Nature* **455**, 627–632 (2008).

C. Zimmer, *She Has Her Mother's Laugh*. Picador, London, 2018a.

C. Zimmer, "Carl Zimmer's Game of Genomes," *STAT News* 2018b. https://www.statnews.com/feature/game-of-genomes/season-one/

PICTURE CREDITS

INDEX